Maintaining Electrotechnical Systems

Malcolm Doughton and John Hooper

CENGAGE
Learning·

Australia • Brazil • Japan • Korea • Mexico • Singapore • Spain • United Kingdom • United States

Maintaining Electrotechnical Systems
Malcolm Doughton and John Hooper

Publishing Director: Linden Harris

Commissioning Editor: Lucy Mills

Development Editor: Claire Napoli

Project Editor: Alison Cooke

Production Controller: Eyvett Davis

Typesetter: S4Carlisle Publishing Services

Cover design: HCT Creative

Text design: Design Deluxe

For product information and technology assistance,
contact **emea.info@cengage.com.**

For permission to use material from this text or product,
and for permission queries,
email **emea.permissions@cengage.com.**

British Library Cataloguing-in-Publication Data
A catalogue record for this book is available from the British Library.

ISBN: 978-1-4080-3999-1

Cengage Learning EMEA
Cheriton House, North Way, Andover, Hampshire, SP10 5BE,
United Kingdom

Cengage Learning products are represented in Canada by Nelson Education Ltd.

For your lifelong learning solutions, visit **www.cengage.co.uk**

Purchase your next print book, e-book or e-chapter at
www.cengagebrain.com

Printed in Malta by Melita Press
1 2 3 4 5 6 7 8 9 10 – 15 14 13

Dedication

This series of study books is dedicated to the memory of Ted Stocks whose original concept, and his publication of the first open learning material specifically for electrical installation courses, forms the basis for these publications. His contribution to training has been an inspiration and formed a solid base for many electricians practising their craft today.

The Electrical Installation Series

Legislation: Health and
Safety & Environmental

Organizing and Managing
the Work Environment

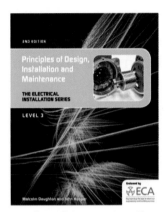

Principles of Design Installation
and Maintenance

Installing Wiring Systems

Planning and Selection for
Electrical Systems

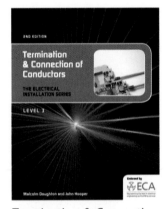

Termination & Connection
of Conductors

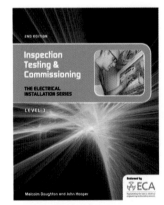

Inspection Testing
& Commissioning

Fault Finding &
Diagnosis

Contents

Acknowledgements

The authors and publisher would like to thank Chris Cox and Charles Duncan for their considerable contribution in bringing this series of study books to publication. We extend our grateful thanks for their unstinting patience and support throughout this process.

The authors and publisher would also like to thank the following for providing pictures for the book:

Belkin Ltd.
BSI
Consort Equipment Products Ltd.
Cooper Lighting and Safety Ltd.
Dehn (UK) Ltd.
Draper Tools Ltd.
Faithful Tools
Hager
HSE
Ideal Industries
The IET

Indigo Plc
Kewtech
Legrand Electric Ltd.
Lifting Gear Direct Ltd.
Malco Tools
Marshall Tufflex
Martindale Electric
Master Lock Europe
Megger
MK Electric
NHP Electrical Engineering Products Pty Ltd.

Pafbag
Peglar Yorkshire Group Ltd.
Protecta Screen Ltd.
Roughneck
Screwfix
Secatol
TLC Direct
Western Disposal Services
Wilmar Corp
Witels Albert USA
www.rubbermaidcommercial.com

Every effort has been made to contact the copyright holders.

This book is endorsed by:

Representing the best in electrical engineering and building services

Founded in 1901, the Electrical Contractors' Association (ECA) is the UK's leading trade association representing the interests of contractors who design, install, inspect, test and maintain electrical and electronic equipment and services.

www.eca.co.uk

About the authors

Malcolm Doughton

Malcolm Doughton, I.Eng, MIET, LCG, has experience in all aspects of electrical contracting and has provided training to heavy current electrical engineering to HNC level. He currently provides training on all aspects of electrical installations, inspection, testing, and certification, health and safety, PAT and solar photovoltaic installations. In addition, Malcolm provides numerous technical articles and is currently managing director of an electrical consultancy and training company.

John Hooper

John Hooper spent many years teaching a diverse range of electrical and electronic subjects from craft level up to foundation degree level. Subjects taught include: Electrical Technology, Engineering Maths, Instrumentation, P.L.C.s, Digital, Power and Microelectronic Systems. John has also taught various electrical engineering subjects at both Toyota and JCB. Prior to lecturing in further and higher education he had a varied career in both electrical engineering and electrical installations.

Study guide

This study book has been written and compiled to help you gain the maximum benefit from the material contained in it. You will find prompts for various activities all the way through the study book. These are designed to help you ensure you have understood the subject and keep you involved with the material.

Where you see 'Sid' as you work through the study book, he is there to help you and the activity 'Sid' will indicate what it is you are expected to do next.

Task

Look up the requirements of the Defence Clause (29) of The Electricity at Work Regulations before continuing with this chapter.

Task A 'Task' is an activity that may take you away from the book to do further research either from other material or to complete a practical task. For these tasks you are given the opportunity to ask colleagues at work or your tutor at college questions about practical aspects of the subject. There are also tasks where you may be required to use manufacturers' catalogues to look up your answer. These are all important and will help your understanding of the subject.

Try this

For star connected windings calculate the line to line voltages if the line to neutral voltages are:

400V _____

220V _____

120V _____

Try this A 'Try this' is an opportunity for you to complete an exercise based on what you have just read, or to complete a mathematical problem based on one that has been shown as an example.

Remember

An isolator must cut off an electrical installation, or a part of it, from every source of electrical energy.

Remember A 'Remember' box highlights key information or helpful hints.

RECAP & SELF ASSESSMENT

Circle the correct answers.

1 The Provision and Use of Work Equipment Regulations (PUWER) applies to:

a. Any equipment provided by an employer
b. An electrician's personal hand tools
c. Only specialist equipment
d. Only electrical equipment

2 An employer's responsibility to provide a safe working environment will not include the provision of:

a. Adequate access to the worksite
b. Suitable and adequate lighting
c. A clean and tidy workplace
d. Operative's hand tools

Recap & Self Assessment At the beginning of all the chapters, except the first, you will be asked questions to recap what you learned in the previous chapter. At the end of each chapter you will find multichoice questions to test your knowledge of the chapter you have just completed.

Note

Further information on renewable energy options is given in the Legislation study book in this series.

Note 'Notes' provide you with useful information and points of reference for further information and material.

This study book has been divided into Parts, each of which may be suitable as one lesson in the classroom situation. If you are using the study book for self tuition then try to limit yourself to between 1 hour and 2 hours before you take a break. Try to end each lesson or self study session on a Task, Try this or the Self Assessment Questions.

When you resume your study go over this same piece of work before you start a new topic.

Where answers have to be calculated you will find the answers to the questions at the back of this book, but before you look at them check that you have read and understood the question and written the answer you intended to. All of your working out should be shown.

At the back of the book you will also find a glossary of terms which have been used in the book.

At the end of Chapter 4, when you have covered the material required for unit 9a, there is an 'End test' and a further 'End test' after chapter 9 covers the material in unit 9.

There may be occasions where topics are repeated in more than one book. This is required by the scheme as each unit must stand alone and can be undertaken in any order. It can be particularly noticeable in health and safety related topics. Where this occurs, read the material through to ensure that you know and understand it and attempt any questions contained in the relevant section.

You may need to have available for reference current copies of legislation and guidance material mentioned in this book. Read the appropriate sections of these documents and remember to be on the lookout for any amendments or updates to them.

Your safety is of paramount importance. You are expected to adhere at all times to current regulations, recommendations and guidelines for health and safety.

Maintaining Electrotechnical Systems

Material contained in this unit covers the knowledge requirement for C&G Unit No. 2357-309a and 9 (ELTK 09a and 9), and the EAL unit QELTK3/007.

This workbook considers the practices and procedures for planning and preparing to maintain, and the maintenance of, electrotechnical systems and equipment. It considers the principles, regulatory requirements and procedures for preparing work sites for maintenance activities (in unit 9a) and for completing the maintenance on electrotechnical systems and equipment (unit 9). It also covers the procedures and documentary systems which underpin work required to maintain electrical systems and equipment.

You could find it useful to look in a library or on-line for copies of the legislation and guidance material mentioned in this study book. Read the appropriate sections and remember to be on the lookout for any amendments or updates to them. You will also need to have access to manufacturers' catalogues.

Before you undertake these units read through the study guide on page ix. If you follow the guide it will enable you to gain the maximum benefit from the material contained in these units.

1

Principles and regulatory requirements for preparing worksites for maintenance

LEARNING OBJECTIVES

On completion of this chapter you should be able to:

- State the appropriate requirements of regulations that are applicable to electrotechnical maintenance work activities, including:

  - Provision and Use of Work Equipment Regulations

  - Electricity at Work Regulations

  - Health and Safety at Work Act

  - Current version of the IEE Wiring Regulations

  - Memorandum of Guidance on the Electricity at Work Regulation 1989.

- Specify the actions required to ensure that electrical maintenance worksites are correctly prepared in terms of health and safety considerations, including:

  - Provision for safe access and exit

- Checking immediate work location for potential hazards as appropriate to property, plant, machinery, personnel and livestock

- Confirm that appropriate risk assessments and method statements have been produced.

- Specify the requirements for preparing and reviewing the work location prior to commencement of maintenance work activities in terms of:

  - Identification of specifications for maintenance, including: drawings, diagrams (circuit and wiring), maintenance schedules/specifications, data charts, manufacturer's manuals, servicing records/running logs, flow charts, standard maintenance time records

  - Organization of a work plan, including: definition of task; planned shut downs/isolations; safety precautions (provision for release of stored and latent energy); permits to work, organizing tools, equipment and spare parts; documentation; communication with relevant parties; time/cost effectiveness

  - Identification and selection of safe isolation methods for: electrical systems and pressurized systems (i.e. hydraulic, compressed air, water, gas)

  - Identification and selection of methods to safely secure work areas including: fences, barriers, screens and warning signs

  - Identification and selection of suitable: hand and power tools (110 V ac or battery operated); portable and fixed lifting equipment; access equipment

  - Provision for safe storage of tools, equipment and materials.

- Identify Personal Protective Equipment appropriate to the work activity being carried out.

- Confirm that tools and equipment are fit for purpose and (where appropriate) are correctly calibrated.

This chapter identifies the principles and regulatory requirements which apply to the preparation of maintenance worksites. As you work through this chapter you will need to have access to the listed regulatory documents and manufacturer's or supplier's information. This reference material may be either as a hard copy or accessed on-line.

Part 1 Relevant regulations

To begin we are going to consider the regulations and requirements that are applicable to electro-technical maintenance work. These include:

- Health and Safety at Work Act
- Provision and Use of Work Equipment Regulations
- Electricity at Work Regulations
- Current version of the IEE Wiring Regulations
- Memorandum of Guidance on the Electricity at Work Regulation 1989.

Health and Safety at Work (etc.) Act

This Act applies to everyone who is at work. The aim of the Act is to:

- Improve or maintain the standards of health, safety and welfare of all those at work
- Protect other persons against risks arising from the activities of those at work
- Control the keeping and use of dangerous substances, for example substances that could be explosive or flammable.

Provision and Use of Work Equipment Regulations (PUWER)

PUWER was introduced under the Health and Safety at Work (etc.) Act 1974 and concerns equipment used at work. Generally, **any** equipment used by an employee during their work is covered.

These Regulations require that the equipment supplied by an employer is suitable for its intended use and that it is safe for use. One particular requirement in respect to maintenance is that it is maintained in a safe condition and inspected to make sure that it remains safe so that people's health and safety is not at risk.

All equipment should be inspected:

- After installation
- Prior to use
- At appropriate intervals and
- If any conditions change, for example if a fault has occurred.

Equipment should only be used and inspected by persons who have been adequately trained, are competent, and have been given instruction in its use. After an inspection has taken place records should be made and kept for the next inspection.

The Regulations state that any risks that are created by the use of the equipment should be controlled by such means as protection devices, guards, warning devices and personal protective equipment (PPE). Maintenance should only take place when equipment is shut down.

The Health and Safety Executive (HSE) produce an Approved Code of Practice (CoP) on the Provision and Use of Work Equipment Regulations 1998 entitled Safe use of work equipment (Figure 1.1). This CoP can be downloaded free of charge from the HSE website.

Figure 1.1 *HSE Code of Practice – Safe use of work equipment*

Electricity at Work Regulations 1989

The Electricity at Work Regulations are statutory regulations which come under the Health and Safety at Work (etc.) Act and are particularly relevant to the electrical installation and maintenance industries.

They require that every employer, self-employed person and every employee comply with safe working procedures to ensure electrical safety in the workplace.

Employers are required to provide and maintain electrical equipment, give training, information and supervision to employees where it is required.

Regulation 4 requires that

'All electrical systems, so far as is reasonably practicable, be of safe construction and maintained in that state.'

Employees are required to take reasonable care for their own health and safety and not endanger others, they must co-operate with their employer on health and safety procedures, not interfere with tools, equipment, etc. provided for their health, safety and welfare and correctly use all work items provided in accordance with instructions and training given to them.

The Electricity at Work Regulations can be downloaded free of charge from the government legislation.gov.uk website.

BS 7671

BS 7671, the Requirements for Electrical Installations, is published by the Institution of Engineering and Technology (IET) and is commonly referred to as the Wiring Regulations.

Whilst these regulations are not statutory they are accepted as standard practice for electrical installation work. They may be referenced in a court of law should prosecution be brought under the statutory regulations.

BS 7671 is divided into seven parts:

Part 1 – Scope, object and fundamental principles

Part 2 – Definitions

Part 3 – Assessment of general characteristics

Part 4 – Protection for safety

Part 5 – Selection and erection of equipment

Part 6 – Inspection and testing

Part 7 – Special installations or locations.

Task

Look up the requirements of the Defence Clause (29) of The Electricity at Work Regulations before continuing with this chapter.

Memorandum of Guidance on the Electricity at Work Regulations 1989

Figure 1.2 *Memorandum of Guidance on the Electricity at Work Regulations 1989*

The Memorandum of Guidance on the Electricity at Work Regulation 1989 (Figure 1.2) gives technical and legal guidance on the Electricity at Work Regulations to all those involved in the design, construction, operation or maintenance of electrical systems. It is designed to help achieve high standards of electrical safety.

The Memorandum of Guidance on the Electricity at Work Regulations 1989 is available as a free download from www.hse.gov.uk.

Task

Familiarize yourself with the Memorandum of Guidance on the Electricity at Work Regulations, in particular, Regulation 4 (2) which is a reference to maintenance work being required to ensure safety of the system.

Try this

Using the Memorandum of Guidance on the Electricity at Work Regulations 1989:

1 Who does the Electricity at Work Regulations Act apply to?

_____

2 The purpose of the Regulations is to prevent death or personal injury when working with electricity. There are five examples given. List them here.

a _____

b _____

c _____

3 a Who judges the frequency with which the system should be maintained?

_____

b On what information is this judgement based?

i _____

ii _____

Part 2 Site preparation

Before maintenance can be carried out the site must be correctly prepared in terms of the health and safety considerations.

Remember

Employers and employees are required, by law, to observe safe working practices.

The employer's responsibility to provide a safe working environment includes such requirements as:

- Maintaining a reasonable working temperature and humidity
- Adequate ventilation and fume and dust control
- Suitable and adequate lighting
- A clean and tidy workplace
- Adequate access and exit
- Suitable sanitary and welfare facilities.

The Management of Health & Safety at Work Regulations requires employers to provide and maintain a working environment for their employees which is, as far as practicable, safe and without risk to health. The term working environment applies to all areas to which employees have access. For example, corridors, staircases and fire exits are included, as are gangways and steps (Figure 1.3).

Figure 1.3 *Access and gangways to be kept clear*

If other employers or contractors are sharing the workplace then the employer is responsible for working with them so that everyone's health and safety is safeguarded.

Other facilities that are required by law include those for washing and sanitation and these must also be suitable for disabled employees.

Preparing the worksite

Electrical maintenance work is carried out in many different locations: domestic, agricultural, commercial and industrial. These premises are often occupied and in use or they may be vacant.

The maintenance electrician's working environment will also vary greatly including:

● A comfortable home
● Damp and noxious atmospheres
● Heat, such as in a boiler room
● Cold, such as in a cold store, or working outside in wintry conditions.

Figure 1.4 *All the comforts of home?*

Risks

Before any work is undertaken the site must be inspected for any potential dangers.

These could include:

● Insecure or unsafe structures
● Inadequate lighting
● The risk of falling or being hit by falling or moving objects
● Risk of drowning
● Dangerous or unhealthy atmospheres
● Steam, smoke or vapours.

Figure 1.5 *Is the structure safe?*

Wherever possible any risks should be removed, or access to the danger area prevented by barriers and warning notices. Consideration must also be given to the provision, wearing and use of the appropriate PPE which includes protective clothing and equipment.

Note

Guidance on the requirements for PPE can be found in the HSE Guidance INDG174 (rev1), A short guide to the Personal Protective Equipment at Work Regulations 1992.

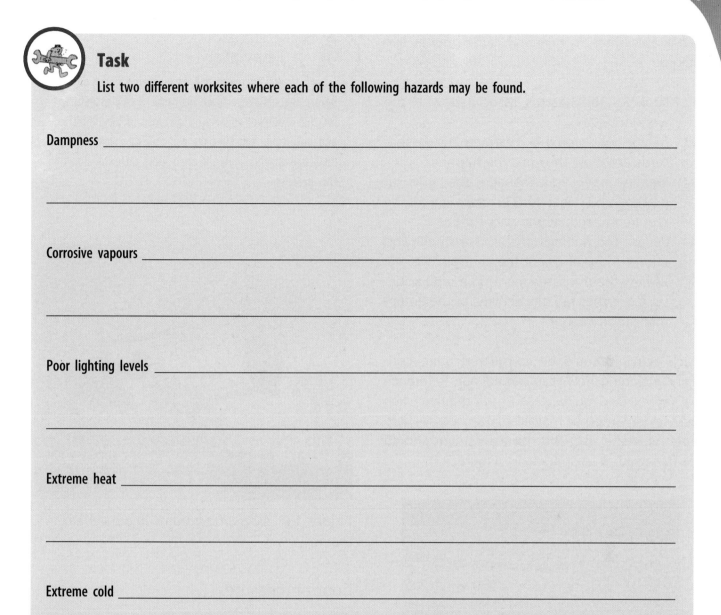

Task

List two different worksites where each of the following hazards may be found.

Dampness _____

_____

Corrosive vapours _____

_____

Poor lighting levels _____

_____

Extreme heat _____

_____

Extreme cold _____

_____

Risk assessments

Before any work is carried out the site must be inspected for potential dangers and one of the first activities is to carry out a risk assessment (Figure 1.6). This should be carried out initially by the employer to ensure the working environment is safe for both the employees and others in the vicinity. In addition we have a responsibility to carry out a risk assessment ourselves as the conditions on-site change and we need to confirm that the worksite is indeed safe.

Figure 1.6 *Inspect for change in conditions on-site*

A risk assessment is generally carried out in five stages:

1   Look for the hazards associated with the work activities
2   Decide who could be harmed by the hazards and how they may be harmed
3   Identify how you manage the risks at present and what further steps might be required to reduce the risks further
4   Record the findings of the assessment and inform those at risk of the controls
5   Review the risk assessment on a regular basis, e.g. if the staff, the activity, or the equipment used changes.

Each activity should be considered quite critically, including how it is carried out. Reference to any existing guidance and information such as accident reports may also be used to help determine the risk. The risk assessment should be reviewed at least annually (Figure 1.7).

Figure 1.7 *Risk assessments are recorded*

Remember

We all make risk assessments every day. Before we cross the road we make a risk assessment for the presence, speed and distance of any vehicles and determine whether we are able to cross safely. Risk assessments at work are simply an extension of this process.

Figure 1.8 *Assess the hazards and determine the risk*

Control measures

Control measures are the ways in which we can reduce the risks. To decide on the most appropriate measures we would consider the following points:

● Can the risk be eliminated altogether?

  ● During our electrical maintenance work a number of the risks may be eliminated if it is possible to isolate the equipment, circuit or installation from the supply.

- Can the risk be contained by additional procedures or equipment?

  - For example, installing conduit at high level could be carried out using a number of methods, such as purpose-built scaffolding or mobile work platforms. Any other associated risks with these systems would also need to be considered such as suitable access, safety harnesses and the need for training.

- Can the work be adapted or arranged to suit a particular situation?

  - Where work is to be carried out in areas where high levels of noise exist, such as in a metalwork shop, one consideration would be to arrange for the work to be carried out whilst there are no operations being conducted that are producing high levels of noise.

- Is there a technological or engineering alternative to carrying out the work?

- One example of this is the use of a remote control robotic camera to carry out surveys in areas which are particularly hazardous such as high temperature or explosive risk areas.

Method statements

Method statements identify the way in which a particular activity is to be carried out and are required for most maintenance activities.

The purpose of the method statement is to detail:

- Precisely what is to be done
- How it is to be done
- Any special requirements or actions
- The time anticipated for the work to be carried out.

Most companies have their own format for producing a method statement, but in general they all contain the above details.

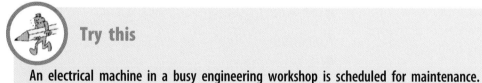 **Try this**

An electrical machine in a busy engineering workshop is scheduled for maintenance.

1 State three possible dangers which must be considered when carrying out a risk assessment.

a _____

b _____

c _____

2 List three significant actions which will need to be identified in the method statement.

a _____

b _____

c _____

Part 3 Preparing for maintenance

Every maintenance task will require information of some kind to allow it to be carried out safely and correctly. This will include such items as drawings, diagrams (circuit and wiring), maintenance schedules/specifications, data charts, manufacturer's manuals, servicing records/running logs, flow charts and standard maintenance time records. The maintenance activities undertaken and any test results will also need to be recorded.

When maintenance takes place on an electrical installation then any certification, including the Electrical Installation Certificate, Minor Electrical Installation Works Certificates and any Electrical Installation Condition Reports (EICR) (formerly periodic inspection reports (PIR)) should be available. These are necessary to identify the isolation and switching arrangements, the circuit identification, protective devices and so on. These documents will also provide evidence of the condition of the electrical installation over time and help to identify any deterioration that may have taken place.

Drawings and diagrams

The most frequently used diagrams that we will come across include:

- Block diagrams
- Circuit diagrams
- Wiring diagrams
- Layout drawings.

Block diagrams indicate the sequence of components or equipment. Each item is represented by a labelled block.

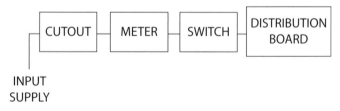

INPUT
SUPPLY

Figure 1.9 *Block diagram*

Circuit diagrams are used to show how the components of a circuit are connected together. The symbols represent items of equipment or apparatus and the diagram will show how the circuit works.

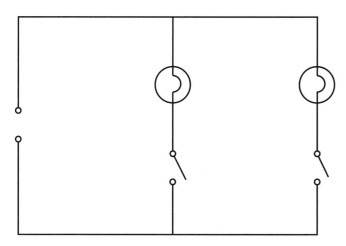

Figure 1.10 *Simple circuit diagram*

Wiring diagrams indicate the locations of the components in relation to one another and cable connections and these are more detailed than circuit diagrams.

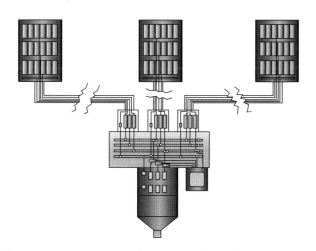

Figure 1.11 *Wiring diagram of intake equipment*

A layout drawing shows the layout of the equipment and the routes of the cables to be installed.

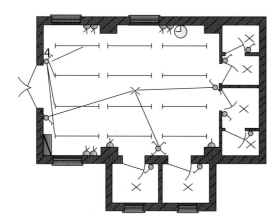

Figure 1.12 *Layout drawing*

Maintenance schedules and records

Maintenance schedules for equipment are usually provided by the manufacturer and should always be followed. Companies may also produce their own maintenance schedules. These will all contain information regarding the interval between visual checks, inspections, testing and overhaul.

Maintenance of Electrical Systems							
	Frequency						
Procedure description	Weekly	Monthly	Bi-monthly	Quarterly	Annually	Bi-Annually	Tri-annually
Emergency Lighting							
Operate test buttons on emergency light fittings	✗						
Operate emergency lighting batteries for 90 minutes					✗		
Main Electrical Switchgear							
Inspect switchgear connections (thermographic)					✗		
Visually inspect for damage and deterioration					✗		
Electrical Switchgear and Distribution Boards							
Visually inspect for damage and deterioration					✗		
Check tightness of connections					✗		
Power shutdown and clean interior							✗
Confirm condition of circuit breakers (visual)							

Figure 1.13 *Sample maintenance schedule for an electrical installation*

The maintenance records should contain information which demonstrates that the schedule is being adhered to and records any measurements taken during the maintenance. This recording of data is important as it determines the trend in performance and provides planning for the next maintenance. It also provides evidence that the system is being maintained for compliance with the statutory regulations.

Task

Every electrician will have their own hand tools and these should be maintained. List five items which should be checked as part of the maintenance for a pair of insulated electrician's pliers.

1 _____

2 _____

3 _____

4 _____

5 _____

Data charts

Data charts show data information in a tabulated form which is readily interpreted visually. Some of the most common are shown in Figures 1.14 and 1.15.

Remember

Tables, charts and diagrams are widely used to give a clear picture of statistical information.

Horizontal bar charts are the most common type of chart used to represent activities and progress for maintenance work.

Electrical Maintenance for Line 2 (Year 3)				
Operation description	Day number			
	1	2	3	4
Pre-work survey	▓			
Visual inspection		▓▓		
Clean and replace contacts/brushes			▓▓	
Electrical testing				▓
Re-commission				▓▓

Figure 1.15 *Typical Bar Chart for electrical maintenance on a production line*

Flow charts

Flow charts are a representation of a set of instructions which must be followed and they are

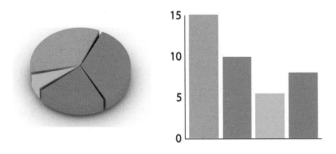

Figure 1.14 *Typical pie chart and vertical bar chart*

made up of different boxes, each of which represents a different function or activity. An example of a flow chart for the safe isolation procedure is given in Chapter 5.

Manufacturer's manuals

Manufacturers' information includes the technical specification for materials and equipment, such as current rating or power, environmental suitability and appearance.

The maintenance requirement for the installation equipment is also included in the particular manufacturer's instructions. The equipment should be maintained in accordance with these instructions. This is in order to ensure that the equipment operates effectively and efficiently. Incorrect maintenance will not only affect the performance of the equipment it will generally invalidate the manufacturer's warranty.

The manufacturer's instructions should be referred to when the equipment is being commissioned and then passed to the client. These should then be made available to those responsible for the operation and maintenance of the equipment once it is in service.

These requirements will also extend to the tools and any equipment used in the maintenance process.

The manufacturer's information usually contains:

- General information on the product such as colour, use, safety information and other useful information for the customer (although this may only be included in their brochure)
- Technical information such as power requirements, current and voltage ratings, physical sizes, mounting positions and any other relevant technical detail
- Installation and commissioning instructions for the equipment, including wiring and lay-out diagrams

- Setting and adjustment information for correct operation
- Maintenance requirements such as the type and frequency of maintenance
- Fault finding charts
- Spare parts information and references.

Of course the extent of the information provided will depend upon the equipment concerned. For example the information provided with a 13 A socket outlet will be considerably less than that provided with an electric shower or a gas fired boiler.

The manufacturer's information may be obtained from their own catalogues, wholesalers' catalogues, brochures and advertising literature or the manufacturer's website. A number of wholesalers and suppliers also provide on-line details for all kinds of equipment, materials and tools.

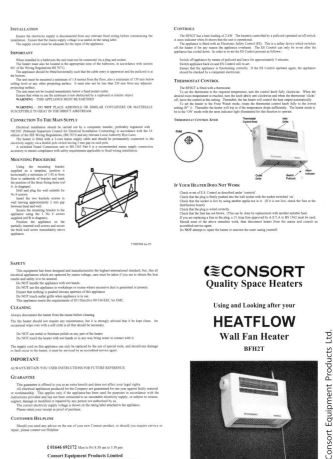

Figure 1.16 *Typical manufacturer's instructions for small appliance*

Records

Before commencing maintenance work the appropriate records should be available and that will generally include:

- Service records; these contain a documentary history of any servicing on equipment or systems
- Running logs; these are a record of events, activities, data, etc. including the dates and signatures of the person carrying out the work
- Maintenance time records; these record the maintenance work carried out on the system or equipment and should give the time intervals required for each item.

Organizing a work plan

Organizing the work plan is a significant part of any planned maintenance work. In order to plan the work we need to consider the following requirements:

1   What system and/or equipment is to be maintained, the extent of the maintenance and the equipment involved is an essential starting point

2   Start and finish dates; the allotted time period for the work will determine the level of labour, tools and materials that may be required

3   Permit to work; maintenance work is often carried out in premises where the normal work activities continue. In such locations a permit to work is often required to ensure that the maintenance work is carried out safely and effectively

4   Planned shutdowns; it is necessary to shut down systems and equipment to carry out maintenance and these periods need to be determined to ensure safety and minimize disruption. This information will be included in the permit to work and will be agreed with the client

5   Safety precautions; in addition to the considerations for electrical safety we need to identify and safely isolate other pressurized systems such as gas, water, compressed air and hydraulic systems

6   Safe isolation; of an electrical installation, circuit or item of equipment is an essential part of the electrical maintenance work. The safety precautions must also consider the discharge of stored and latent energy. In the electrical system this is stored in equipment such as capacitors, ac machines and solar panels

7   Tools, equipment and spare parts required; to ensure that the maintenance work is carried out safely and in the allotted time it is important to make sure that the necessary tools, equipment and any spare parts are available before the work begins

8   Relevant documentation; the necessary documents, as discussed earlier, should be available to the person carrying out the maintenance, prior to work commencing. Some of this information may be needed earlier to allow spares to be ordered

9   Any relevant parties involved; we need to establish any other persons involved in or affected by the maintenance activities in order to keep them informed on the activities and progress

10 Time and cost effectiveness; when planning the maintenance work the effect on the client and the activities of others is a prime consideration. The work needs to be planned to ensure that there is the minimum of disruption. The timing and duration of the work should also be considered to provide the most cost-effective solution.

Task

Go on-line and look at a typical annual service schedule for an air conditioning unit and in particular note the electrical checks that are required.

_____

_____

_____

_____

_____

_____

_____

_____

_____

_____

_____

_____

_____

_____

Isolation of electrical services

In order to work safely on equipment which has previously been put into use it is essential to ensure the equipment has been isolated before work commences. This isolation relates to the electrical installation and equipment and other services. We need to consider the devices which may be used to carry out this isolation before we continue.

Remember

The information detailing the isolation, switching and control of the installation and equipment is essential for maintenance work to be carried out safely.

We will begin by considering the equipment which provides electrical isolation.

BS 7671 defines isolation as:

'A function intended to cut off for reasons of safety the supply from all, or a discrete section, of the installation by separating the installation or section from every source of electrical energy.'

The requirements of the Electricity at Work Regulations (EWR) must be followed when isolating the supply.

Before any circuit or equipment is worked on it should be isolated from the supply. Many accidents have occurred where the wrong circuit has been isolated. It is therefore very important that all circuits and isolators are clearly labelled to identify what it is they control.

Figure 1.17 *Immersion heater isolated by a double pole switch*

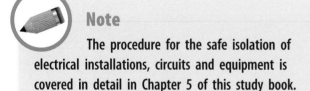

Note

The procedure for the safe isolation of electrical installations, circuits and equipment is covered in detail in Chapter 5 of this study book.

Circuits and equipment should be tested to confirm that they are isolated before working on them. It is important that the isolator cuts off the electrical installation, or any part of it, from every source of electrical energy.

The device used to isolate the supply must not only be suitable for operation in both normal and abnormal conditions but also be placed in an accessible location. Where switches are used as isolators a clear air gap must exist between the contacts so that they cannot accidentally reconnect. They must also be clearly marked so that there is no doubt whether the switch is on or off.

Some typical examples of suitable isolating devices can be seen in Figures 1.17 to 1.21.

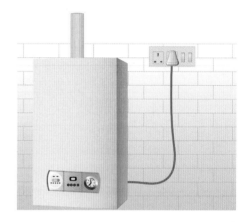

Figure 1.18 *Domestic boiler with isolation by use of a plug and socket*

Figure 1.19 *Three phase distribution board with triple pole and neutral isolator*

Figure 1.20 *Motor with adjacent isolator*

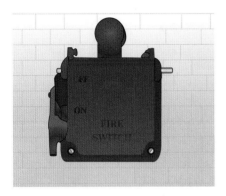

Figure 1.21 *Fireman's switch emergency isolator*

Once the equipment has been isolated to work on it there are a number of actions which need to be taken including:

● All exposed electrical connections should be tested to see if they are dead

● The test equipment used to confirm isolation must be proved before and after use

● The switch or isolator should be locked in the off position

● Notices posted at the isolator warning that the equipment is being worked on.

The Electricity at Work Regulations state that work on or near live equipment should be carried out only when:

● It is unreasonable in all circumstances for it to be dead

● It is reasonable in all circumstances to work on or near the live equipment

● Suitable precautions are taken including suitable procedures and protective equipment to prevent injury.

If it becomes essential to work on or near live equipment a permit to work may be necessary. This ensures that both the person working on the live equipment and a person in authority know of the risks that are being taken.

Work on live equipment must never be undertaken without first examining every other possibility. Should it be necessary to work live, every safety precaution must be considered and taken to prevent danger.

Once isolation has been carried out and before work can be started the conductors or equipment must be proved dead. An approved voltage indicator which conforms to the Health and Safety Executive Guidance Note GS 38 is recommended for this and must be checked before and after testing to ensure that they are working correctly. This is usually carried out with the aid of a proving unit to confirm the approved voltage indicator is functioning before and after testing.

Courtesy of Kewtech

Figure 1.22 *Typical approved voltage indicator and proving unit*

Where the electrical equipment has its own source of supply, such as batteries, capacitors or generators contained within them these generally cannot be isolated. It is important that in such circumstances all possible precautions are taken to prevent a dangerous situation occurring.

Warning notices should be posted where equipment is isolated warning that work is being carried out on the equipment and that it would be dangerous to reconnect it.

Figure 1.23 *Heavy duty lockout tags*

If there is any live equipment adjacent to isolated equipment then danger notices should be posted to advise these items are live. Once work is completed these notices must be removed.

Isolation of other services

Other systems such as hydraulics, compressed air, water and gas may also need to be isolated to carry out maintenance safely. These systems incorporate isolating valves for just such a purpose.

Figure 1.24 *Services lockout devices*

The basic premise is the same for these services as it is for the electrical equipment. Ensure the isolating valve is in the off position, using a proprietary lockout device, lock the valve in the off position and retain the key. Post-warning notices to advise that the service has been deliberately isolated and work is being carried out. The notices shown in Figure 1.23 may be used for this purpose.

Figure 1.25 *Isolating valves a) Gate, b) Isolating, c) T-ball valve, d) ball*

A proprietary lock off kit for maintenance engineers is a worthwhile investment. These contain both electrical and other service lock off devices and the necessary keys, padlocks and labels.

Figure 1.26 *Proprietary maintenance lockout kit*

Once the supply is isolated then the system will need to be carefully bled to discharge the pressure and content of the system downstream of the isolation valve.

Task

Using manufacturer's or supplier's details select a suitable lock off device for each of the following and record the catalogue description and number.

1 A BS EN 60898 circuit breaker

2 A small gate valve

3 A ball valve

Try this

List the requirements identified in the Electricity at Work Regulations which must be met before it is considered reasonable to work on or near live equipment.

1 _____

_____

2 _____

_____

3 _____

_____

Securing the work area

Safely isolating and making the system safe to work on is a major part of the safety requirements for maintenance work. However we also have to make the area in which the work is to be carried out safe for us and other users of the premises.

Where live testing has to be undertaken accidents often occur due to distractions or interference from other people. Where testing is to be carried out and access to live terminals is necessary there can also be a danger to other people in the vicinity.

It is important that we carry out a risk assessment and take the necessary precautions to prevent danger to ourselves and other people. Barriers and warning signs will be necessary to advise people of the work activity and keep them away from danger.

Barriers and warnings will be needed where the work involves:

● Having open access to holes in floors, service ducts, etc.
● Using steps or ladders
● Activities in thoroughfares such as corridors, access routes and stairwells
● Working on or near live conductors or equipment.

People often ignore or "test out" the barriered off areas and warning notices, and so these may not be sufficient in some areas. Where maintenance work is to be carried out in busy areas it may be necessary to use fencing or screens to prevent interference from other people and to ensure their safety. Outdoor areas often require the use of portable fencing to ensure that the work area is kept free. Screens may be used to prevent

people congregating to watch what is going on or to prevent dust and materials drifting into other areas.

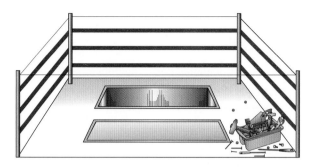

Figure 1.27 *Place barriers around open floor access points and drops*

Whenever maintenance work is being carried out there will be on-site checks that need to be made to ensure safety including:

● The site is kept tidy, and materials are stored safely
● Operatives can reach their place of work safely, e.g. roadways, gangways, passageways, staircases, ladders, scaffolds and passenger hoists are accessible and in good condition
● All holes, openings and excavations are provided with secure guard rails or with fixed, clearly marked covers and barriers to prevent falls
● There are guard rails fitted to stop falls from open edges on scaffolds, mobile elevating work platforms, buildings, gangways and excavations
● Suitable safety harness attachment points are provided
● The building work structures are secure, stable and adequately supported
● All working areas and access walkways are free from trip hazards and obstructions such as stored materials and waste
● Proper arrangements for collecting and disposing of waste materials are in place and maintained

- The work area, both external and internal, is adequately lit
- Sufficient local lighting is provided where work is carried out after dark or inside buildings.

In addition there would be a need to confirm trade-specific requirements such as whether suitable reduced low voltage supplies are available for power tools etc.

Figure 1.28 *Safety barrier*

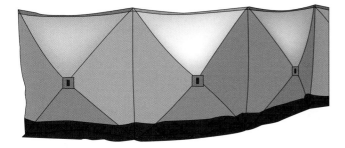

Figure 1.29 *Safety screen*

Selecting suitable tools, lifting and access equipment

Employers are required to provide and maintain suitable safety tools and equipment for use by their employees. Training in the use of such equipment, where this is necessary, must also be provided. Equipment should be inspected pe-

riodically to ensure that it remains safe for use and records should be kept of the inspections carried out on the equipment.

Figure 1.30 *Rotary hammer*

For our maintenance activities the type and suitability of the tool and equipment must be considered. If the work requires power tools, are reduced 110 V ac tools required or can battery operated tools be used? The use of 110 V ac power tools may only be required where there is a lot of heavy duty drilling, chasing or cutting involved. Battery powered tools are suitable for many of the tasks and do not require a power source at the workface, have no trailing leads and are particularly useful when the electrical supply there is isolated.

Handling and transporting materials is also the responsibility of the employer and heavy items may need to have mechanical handling aids, such as a sack barrow, pallet truck or fork-lift truck, for their safe transportation.

Figure 1.31 *A fork-lift truck*

For some of the maintenance activities access equipment will be necessary. This can range from a simple work platform to a full scaffolding, or a mobile elevated work platform (MEWP).

Safe storage of tools, equipment and materials

When equipment and materials are brought onto the site provision must be made for their safe and secure storage. It is important to protect all the materials and tools and equipment to be used for the maintenance from damage or theft. In many larger organizations there is a storage facility on-site which may be used for the company stock spares. This may be suitable for the storage of the maintenance material and equipment. Where this is not the case the client may have a lockable room or storage cupboard which can be used for the duration of the work.

For some maintenance activities the material and equipment requirements are such that a site hut or store outside the main building may be required. This may be in the company car park for example. This storage facility must be laid out so that the materials and equipment are not damaged.

The atmosphere in the store should not be allowed to become damp to ensure the equipment remains in good condition prior to installation on-site. Electronic components, replacement lamps, fluorescent luminaires and such like, need to be stored to prevent any possible mechanical damage.

For smaller maintenance activities it may be sufficient to provide a site box which will allow the materials and equipment to be stored close to the work area.

© Secatol

Figure 1.33 *On-site storage box*

Courtesy of Western Disposal Services

Figure 1.32 *Large temporary site storage facility*

Remember
Product quality is preserved and protected by using the correct handling techniques. It is important to ensure that:

● Stock is unpacked using the correct techniques and equipment

● All packing is removed and disposed of promptly and in the correct manner

● Discrepancies and/or damaged stock are set aside to be dealt with correctly

● In adverse conditions precautions are taken to prevent damage.

Task

Using manufacturer's or supplier's details select suitable equipment for each of the following activities and record the catalogue description and number.

1 To secure the area around a cutting machine for maintenance work.

2 To move a pallet of replacement fluorescent lamps through a working office.

3 To store small tools and components at the point of maintenance in a factory.

Personal protective equipment

Personal Protective Equipment (PPE) is all the equipment intended to be worn or used by a person at work and which offers protection against one or more risks to their health or safety.

In terms of dealing with risks at work PPE is a last resort. The risks must be assessed and every other option taken to remove or reduce the risk of danger and injury. It is only if there is no other way of providing the protection necessary to do the work safely that PPE should be supplied.

The main requirement of the Personal Protective Equipment at Work Regulations is that wherever there are risks to health and safety that cannot be adequately controlled by any other means then PPE must be supplied and used.

Figure 1.34 *Work gloves are one type of PPE which may be required*

Employers should consider all means of protection and assess whether PPE is required and where it is the employers are required to ensure that PPE is:

● Properly assessed before use to ensure that it is suitable for the situation, fits the wearer and prevents or adequately controls the risks that may occur

- Kept clean, stored and maintained in good repair
- Provided with instructions for its correct use and
- Used in the proper manner by employees.

Remember

You have a duty to ensure that you know, understand and use the equipment provided correctly. If the equipment is provided it must have been assessed as necessary. Never fail to use the PPE provided because the job will only take a few minutes.

Tools and equipment

Tools and equipment are an essential part of the maintenance electrician's kit and they need to be fit for their purpose. The basic electrician's tools will include cutters, pliers, screwdrivers, saws and the like. Regular checks will need to be made to ensure that these are in good condition and working correctly. Many accidents during maintenance are caused by unsuitable or poorly-maintained hand tools.

Image provided by Screwfix

Figure 1.35 *Typical basic tool kit*

During any electrical maintenance work there is going to be some element of measurement to be done. This may be assessing a simple confirmation of the dimensions of a luminaire or the size of an enclosure using a standard pocket tape. In some cases a cable may need to be replaced and so the cable length will need to be determined and depending on the length and location a long tape or measuring wheel may be necessary.

Electrical testing will need to be undertaken and the test equipment used for this will need to be correctly selected. In order to carry out testing correctly and safely the appropriate test instruments must be used. These must be suitable for the test to be carried out and safe to use. A check on the condition, the current calibration and the function of the instrument, where possible, will need to be made.

The instruments which may be required to carry out the tests involved in the maintenance activity on the electrical installation are:

- Low resistance ohmmeter
- Insulation resistance ohmmeter
- Earth fault loop impedance test instrument
- Prospective fault current tester
- RCD tester
- Phase sequence test instrument
- Voltmeter
- Ammeter.

© Megger

Figure 1.36 *Typical test instruments*

The information provided by the manufacturer for the test instruments details the operating parameters and characteristics of the test instrument. It is important to confirm that the test instruments meet the requirements of BS 7671, Requirements for Electrical Installations, the IET Wiring Regulations and that they are suitable to use. Test instruments used by electricians should also meet the requirements of HSE Guidance GS 38.

Note

The requirements for the test instruments may be found in BS 7671 and the IET Guidance Note 3 Inspection and Testing. The HSE publishes further guidance relating to test instruments used by electricians in the form of HSE Guidance GS 38.

Task

Using manufacturer's or supplier's details select suitable items of PPE equipment for each of the following activities and record the catalogue description and number.

1 Drilling new fixings for a replacement luminaire on a concrete ceiling in a workshop.

_____

_____

_____

2 Moving general waste materials from the worksite to a skip.

_____

_____

_____

3 Maintaining a bench saw in a busy machine shop.

_____

_____

_____

Congratulations you have now finished the first chapter in this study book. Complete the self assessment questions before you continue to the next chapter.

SELF ASSESSMENT

1 The Provision and Use of Work Equipment Regulations (PUWER) applies to:

   a. Any equipment provided by an employer
   b. An electrician's personal hand tools
   c. Only specialist equipment
   d. Only electrical equipment

2 An employer's responsibility to provide a safe working environment will not include the provision of:

   a. Adequate access to the
   b. Suitable and adequate lighting
   c. A clean and tidy workplace
   d. Operative's hand tools

3 The first stage of a risk assessment is to:

   a. Cordon off the work area
   b. Produce a method statement
   c. Ensure the electrical supply is isolated
   d. Check for potential hazards in the work area

4 The safe isolation of a pressurized system should be followed by:

   a. Disconnection of the system
   b. Controlled discharge of pressure
   c. Barriers placed around the work area
   d. Notification to the client and other operatives

5 Figure below shows a:

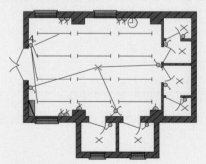

   a. Circuit diagram
   b. Block diagram
   c. Layout drawing
   d. Wiring diagram

Checking the work location

2

RECAP

Before you start work on this chapter, complete the exercise below to ensure that you remember what you learned earlier.

The Health and Safety at Work (etc.) Act is a _____ document and applies to all _____ at _____ .

PUWER requires all equipment to be inspected after _____, before _____ and then at _____ intervals and if any of the _____ or _____ conditions change.

Before any work is carried out the site must be _____ for potential _____ and one of the first activities is to carry out a _____ assessment and the first _____ is; can the risk be _____ altogether?

The ways in which risks can be reduced are called _____ measures. Method statements _____ the way in which an activity _____ be carried out.

The most frequently used diagrams that we will come across include _____, _____ and _____ diagrams.

Maintenance _____ for equipment are usually provided by the _____ and should always be _____ .

The maintenance _____ should contain information which _____ that the _____ is being followed and records any _____ taken during the maintenance.

_____ bar charts are the most common type of chart used to represent the _____ and _____ of maintenance work.

In order to work _____ on equipment which has previously been put into use it is _____ to ensure the equipment has been _____ before work commences.

Once the equipment has been _____ to work on it all electrical connections _____ during the work should be _____ to confirm they are _____.

To safely work on pressurized services, we must make sure the _____ valve is _____ in the _____ position and keep the _____ on our person.

The work must be made _____ for _____ and the _____ users of the _____.

_____ and _____ signs can be used to _____ people of the work _____ and _____ them away from _____.

When equipment and _____ are brought onto the _____ they must be kept in a _____ and _____ store.

Risks must be _____ and every measure taken to _____ or _____ the risk of _____ and injury. _____ is _____ used when there is no _____.

LEARNING OBJECTIVES

On completion of this chapter you should be able to:

● State the preparations that should be completed before electrical maintenance work starts, including:

  – Interpretation of specifications and maintenance schedules to produce accurate material and equipment requisites

  – Identification and selection of material, equipment and components compatible to specification or maintenance schedule

  – Confirmation of site readiness for maintenance work including considerations of building structures and fabric

  – Confirmation that tools, equipment and instruments are fit for purpose

- Confirmation of secure site storage for tools, equipment, materials and components

- Identification of suitable access equipment

- Identification of suitable lifting equipment

- Identification of suitable work methods

- Identification of points in the maintenance programme where coordination with other trades and personnel may be necessary.

● Explain how to check for any pre-existing damage to client property, such as:

- Building wall/floor fabric

- Plant and machinery

- Equipment and components

- Building décor and floor finishes

and state why it is important to do this prior to commencement of any work activity

● State the actions that should be taken if pre-existing damage to customer/client property, plant or machinery is identified

● Specify methods for protecting the fabric and structure of property, plant or machinery before and during maintenance work.

Part 1 Preparations before beginning work

Prior to starting any maintenance work it is important to confirm that everything is in place for the work to go ahead. Once we have an agreed contract of work we must begin the organization and ordering of the materials, plant and labour. The first stage of this process is to correctly identify what is required. This information is obtained from the maintenance programme, the specification and the maintenance schedules. You will find more information on how to do this in the study book 'Organizing and managing the work environment' in this series.

Before we go to the site we need to confirm that the appropriate material, plant and tools have been ordered and are available for use. Not having these in place will result in delays, inconvenience to the client and additional costs.

There are a number of areas which need to be considered in this process and we shall briefly examine these together with the actions that are necessary to ensure suitable preparation for the work to begin.

Materials and equipment

We need to establish a detailed list of material and equipment required to carry out the maintenance work. To do this we need to refer to the maintenance schedule and the manufacturer's maintenance information for the equipment.

For some maintenance activities this may be relatively straight forward. For example the maintenance of the fluorescent lighting in a workshop may require new lamps for the luminaires and a small stock of ballasts and capacitors for any failures. Maintenance on machines and production line equipment may require considerably more varied spare parts which may include belts, motor brushes, contacts and lubricants.

It is important that the maintenance schedule, the manufacturer's maintenance information and the maintenance records are used in this process. Maintaining the equipment is not unlike servicing a car where not every activity is carried out at every scheduled service. Some maintenance will be required at every scheduled maintenance stage while other maintenance activities may occur on every third scheduled maintenance stage and so on.

It is common for organizations to combine the schedule and the record sheet as this provides the maintenance electrician with the schedule as an aide-memoire and provides a record sheet for the maintenance records. A typical example of this is shown in Figure 2.1.

	Air Conditioning Service Record (annual)			
	Task	Action	Results	
1	Fit refrigerant gauges			
2	Check Compressor	Mount kit	Yes / No	
		Tension device	Yes / No	
		Clutch bearing	Yes / No	
		Bolts	Yes / No	
3	Check Belt	Replace belt	Yes / No	
4	Check condenser	Secure	Yes / No	
		Clear of debris	Yes / No	
		Fans operating	Yes / No	
5	Check hoses	Fittings	Yes / No	
		Secure	Yes / No	
		Chafing	Yes / No	
6	Check evaporator	Secure	Yes / No	
		Clear of debris	Yes / No	
		Blower operating	Yes / No	
7	Pressure switch	LP		bar
		MP		bar
		HP		bar
8	Electrical	Fuse Rating		A
		Connections at		
		battery		
		ignition		
		polarity	Correct	
		earth fault loop Z_s		Ω
9	Controller settings	Hi	Yes / No	
		Lo	Yes / No	
		Fan speeds	Yes / No	
		Manual stat operation	Yes / No	
		Re-circ controls operating	Yes / No	
10	Check for leaks	Dye added	Yes / No	
11	System pressure- leak test	25 minutes		bar
12	Reclaim refrigerant		Yes	
13	Replace filter drier		Yes	
14	Evacuate system			
15	Recharge system	Refrigerant type		
		Refrigerant charge		kg
		Oil type		
		Oil added		ml
16	Run and test unit	Pull down to set-point	Yes / No	
17	Additional work			
18	Recommendations	Unit left in service	Yes / No	
19	Engineer details	Name:	Date:	

Figure 2.1 *Typical maintenance schedule and record.*

Once we have identified the materials and spares needed we also have to identify what tools and plant may be necessary beyond the standard electrician's tool kit.

Depending on the maintenance activities there are likely to be some additional tools required. This could include tools such as a torque wrench, feeler gauges, thread gauges, taps and dies in a variety of sizes. We may also need some additional task lighting and access equipment. Materials, tools and equipment which have to remain on-site must be protected from damage or theft. A site hut, or a room on the site which can be locked, may be required to provide secure safe storage.

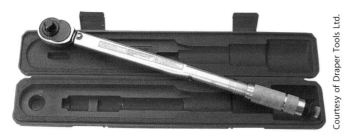

Figure 2.2 *Torque wrench*

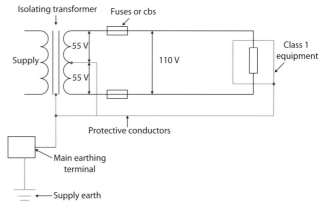

Figure 2.3 *Typical 110 V centre tapped reduced low voltage supply*

During the maintenance work the local electrical supply is normally isolated and therefore we may have to consider the use of battery power tools and local task lighting. Where power is available the use of a centre tapped reduced low voltage supply system should be considered as the preferred option. This system can be used to supply both the power tools and the portable task lighting.

Try this

Using the air conditioning record schedule in Figure 2.1, produce a list of the materials and spare parts which are needed to allow this work to be caried out.

Materials and spares

Product and image by Faithful Tools (www.faithfulltoo s.com)

Figure 2.4 *110 V distribution unit*

There may also be the need for access equipment for the particular tasks being undertaken. This may be a simple low level work platform or a mobile elevated work platform (MEWP). Steps and ladders are permissible for work over short duration and may be suitable for some short period activities such as electrical testing and lamp changing.

Before we actually commence work we will need to confirm both the suitability and availability of the material, equipment and plant that has been ordered. The tools and equipment must also be

Task

You are to carry out the annual maintenance to a refrigeration unit for a butcher's shop. The unit is mounted on the rear wall of the shop at a height of 4 m and the area is used by the local shop owners for deliveries.

Determine, using manufacturers' and suppliers' catalogues, the access and safety equipment required to carry out the maintenance work safely. The tools required for the work should not be included.

Access and safety equipment		
Item	**Quantity**	**Catalogue Reference No.**

fit for the tasks they are being used for. This is generally referred to as being 'fit for purpose'. For example having a set of metric taps on-site when all the machine threads are BS Whitworth will delay the work until the right ones are obtained.

Testing of the electrical systems forms part of the electrical maintenance and may be carried out throughout the maintenance work. It is important to confirm that the test instruments to be used are suitable, including having in-date calibration certificates, are free from damage and function correctly. The leads for the test instruments must also be suitable and meet the requirements for the tests to be carried out.

During maintenance work it may be necessary to use lifting equipment to move or replace components and equipment. This lifting equipment must be suitable for the task in hand and that includes making sure it:

● Is suitable for use within the work environment
● Can safely lift the load
● Will not endanger others during the movement and lifting.

Lifting equipment can be as simple as a hand pumped hydraulic pallet truck and an A-frame lifting support to remove or replace an electric motor. It may require a fork-lift truck to move and raise the equipment into position, such as changing a chiller unit on a cold store flat roof. In extreme cases we may need a full mobile crane and road closure arrangements to lift replacement units onto a multi-storey office building in the city.

Figure 2.5 *Hand pallet truck*

Figure 2.6 *Fork-lift truck*

Lifting replacement solar photovoltaic panels, for example, involves little weight but they are often physically large and easily affected by a light breeze. A suitable lifting mechanism must be used, generally a block and tackle on the working scaffold. Proprietary lifting bags are available to minimize the risk of damage to the panel and allow the lifting process to be controlled using guide ropes.

Lifting Gear Direct Ltd

Figure 2.7 *Mobil A frame hoist*

© Pafbag

Figure 2.8 *Lifting solar panels using block and tackle and lift bag*

The materials and equipment within the installation are normally fixed into position. It is important that we check the security of the fixings and suitability of the building construction. This is particularly relevant where items of equipment are replaced, where the suitability of both the fixing and building structure must be able to support the replacement equipment.

As the maintenance work takes place it is likely that we will need to liaise with other trades. Before work begins we must identify the points in the maintenance programme where we will need to coordinate activities with these other trades. Failure to identify these requirements

before we start work will generally lead to delay incurring additional costs and causing inconvenience and possible loss of revenue to the client.

Careful review of the work program is important to identify such activities. The programme shown in Figure 2.9 shows typical activities in a maintenance programme for a production line. If we are responsible for the electrical maintenance we need to identify activities where we need to coordinate with other trades. To do this we can consider the columns giving details on the description, earliest start and earliest finish.

The first area where coordination will be required is the 1) decommissioning and making safe the work area. At this stage we need to know the electrical requirements of the other trades during their maintenance work for power and lighting. We also need to know whether any part of the production line system will need to be energized during their work.

The second activity where coordination will be required is with the 3) Service and check of the hydraulic system. This overlaps with the 4) Service of motors and control systems on day 3. We would need to check where any work is to be carried out in common areas and coordinate appropriately to ensure the trades do not clash.

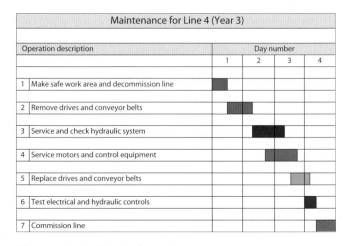

Maintenance for Line 4 (Year 3)				
Operation description	Day number			
	1	2	3	4
1 Make safe work area and decommission line	■			
2 Remove drives and conveyor belts		■		
3 Service and check hydraulic system			■	
4 Service motors and control equipment			■	
5 Replace drives and conveyor belts				■
6 Test electrical and hydraulic controls				■
7 Commission line				■

Figure 2.9 *Typical maintenance programme*

Try this

Using the information in Figure 2.9 highlight all the areas where coordination with other trades may be necessary.

Part 2 Checking the work location

Before we undertake any work it is important to check the condition of the work location. Earlier we considered the requirements for making a risk assessment and so on, but we must also consider the conditions at the time work is to begin. These requirements will vary depending on the type of worksite and the activities being undertaken.

Access to and exit from the site may be affected by production processes and the everyday activities of the client which can then affect the work progress. Carrying out maintenance in a working environment can also cause problems in accessing the work area.

Maintenance in occupied buildings

Electrical maintenance work in occupied buildings and installations introduce some additional considerations. Not least of these is the preparation of the existing site to allow work to commence. Access to and exit from an area should be arranged to allow the work to be undertaken whilst the everyday activities continue with the minimum of disruption. Access routes need to be clearly identified and barriers

and warning notices placed to advise others of the work. It may be necessary to provide protection to the building fabric and install screens and the like to protect the surrounding areas from contamination from dust, dirt, fumes and vapours.

An important part of the preparation for maintenance is to protect the building fabric and minimize the disruption to the occupants. Until this preparation work is in place and is operational maintenance work cannot begin.

Checking for pre-existing damage

It is important to check the client's property for any pre-existing damage. Before we begin work we need to identify any damage to:

- Building fabric, plaster, brickwork and the like
- Finishes including paint, wallpaper, soft furnishings, etc.
- Equipment and appliances
- Carpets, floor coverings and floor finishes
- Ceilings and decoration.

Figure 2.10 *Important to notify and record any existing damage*

The work areas, the access and exit routes and the other areas in the immediate vicinity should be included in the survey. For example maintenance work is to be carried out on the electrical installation in the fitness suite of a hotel. There is a set of double doors for equipment access to the room from the hotel grounds. Before work begins, a check should be made in the work area to include the condition of the:

● Carpets and flooring for tears, stains, scratches, etc.
● Furnishing and fittings within the suite for damage, scratches and other defects
● Doors, door frames, locks and catches, window frames and glass
● Wall coverings and blinds
● Ceiling, lighting and other high level fitments.

In addition checks would need to be carried out along the approach route to the suite including lawns, paving and any ornaments, plants or shrubs and so on. A check would also need to be made in the adjacent rooms and areas to confirm there is no pre-existing damage. If the client complains, after the work has begun, that the wall in the adjacent room has now got dirty marks across it, it would be difficult to rule out the possibility that this is as a result of the maintenance work. This inspection process often brings to light some additional precautions which need to be taken to protect adjacent areas.

Any damage or defects should be brought to the client's attention and recorded. This informs the client of the pre-existing condition and removes the possibility of claims for damage once the work is completed. If we fail to do this there is no evidence to prove that the damage was not caused by our work activities and your company may have to cover the cost of repairs. Such claims can be very costly and will affect the company's insurance premium and may do considerable damage to the company's reputation.

Recording pre-existing damage

If any pre-existing damage is discovered this should be brought to the client's attention immediately and this is generally done verbally if the client is present. This should be followed up with a written report identifying the damage and this should ideally include photographs. These will record for both parties the actual location and extent of the damage.

If the client is not in attendance on-site then it may be possible to telephone the client to advise them of the situation. The written report could then be sent to confirm the telephone conversation.

An acknowledgement from the client that they have been informed of the situation and accept the condition should be obtained before work commences. It is also important to inform the client if it is likely that the work activity could make the existing damage worse or in extreme cases need remedial action before maintenance can be carried out.

Protecting the fabric and structure

Before any maintenance work is undertaken it is important to provide suitable protection for the fabric, structure and fittings within the work area. The type of the protection will depend on the nature of the building fabric and fittings and the type and extent of the work to be undertaken.

Figure 2.11 *Surface protection for fixtures*

Where light work is to be carried out with all the materials hand-carried to the worksite and no heavy equipment is required the protection can be relatively minor. Using dust sheets and plastic floor protectors over the carpets and furniture in an office or dwelling for example.

Distributed in UK & Ireland by: Protect International Ltd.

Figure 2.12 *Protecting an area of the worksite*

However if the work is to be carried out in a property with antique wooden floors for example, then a simple plastic cover or dust sheet may not provide sufficient protection. A felt underlay with hardboard or plywood walkways for access may be required.

Figure 2.13 *Staircase closed to protect other levels from dust*

The use of corrugated plastic sheet is a common practice replacing the plywood protection as it is easier to cut and fit and provides excellent protection qualities. Where heavy impact is likely then the use of plywood is still common.

Figure 2.14 *Corrugated plastic floor protection*

Where large items of equipment and materials are required in an office block it may be that the building lift has to be used to transport men, materials and equipment to the various worksites. In such circumstances the lift should be protected to prevent damage to the lift walls and floors.

Figure 2.15 *Office corridor in the process of being protected*

© Indigo Plc

Figure 2.16 *Door frame protection installation*

Any fine finish doors may need to be removed and stored in a safe location and where this is not practical suitable protection should be provided. Door frames and pillars can be wrapped in protective padding to prevent damage whilst moving materials and equipment. The walls in the main access areas may need to be lined with protective plastic or ply to prevent damage. Clear plastic sheeting can be used to cover windows, taped to the surround to keep it in place.

It is important to identify and notify the client of any pre-existing damage and then to take adequate precautions to prevent any further damage occurring during the work. Failure to take adequate precautions could involve considerable additional expenditure and can seriously damage the company's reputation.

Task

Maintenance is to be carried out on an air conditioning unit located in the suspended ceiling of a company boardroom. The boardroom has an Axminster carpet and a large boardroom table and chairs together with a number of glazed display cases none of which can be moved out of the room.

1   List the areas which would need to be checked for pre-existing damage in the boardroom.

_____

_____

_____

2   Using manufacturer's information list the most suitable method of protection to ensure that no damage occurs during work for the following items:

a   the carpet

_____

_____

_____

_____

_____

b   the table

_____

_____

_____

_____

_____

c   the other furniture within the room

_____

_____

_____

_____

_____

Try this: Crossword

Across

1   To ensure good progress we must ... with other trades (6)

3   Back straight and knees bent for this (7)

6   Clothing material (6)

7   The way in (6)

11 This type of item can be easily carried (8)

12 A safe and secure way for keeping tools (7)

13 Could be a lorry of a lifting device (5)

14 Keeps it safe (10)

16 Hard to let go or a torque ... (6)

17 These are OK for short periods of work (7)

Down

2   Like 17 across only freestanding (5)

4   Raise on prongs (8)

5   This may need to be maintained (9)

8   The type of plastic used for floor protection (10)

9   This could be used in place of 8 down (7)

10 Used to work at heights for longer periods (8)

12 The order of things (8)

15 Hand and special may be needed for maintenance (5)

You have now completed the second chapter of this study book. Successfully complete the self assessment questions before going to the next chapter.

SELF ASSESSMENT

1   To confirm the correct materials are ready to maintain an item of equipment the engineer would check with the:

   a. Manufacturer's instructions

   b. The installer's instructions

   c. Equipment block diagram

   d. Equipment circuit diagram

2   During mantenance within a plant room several trades will be working in the same area. Before work begins it important that the electrical activities in these areas are:

   a. Carried out in sequence

   b. Coordinated with other trades

   c. Completed on a first there basis

   d. Carried out simultaneously

3   During the maintenance of an office building a replacement lighting control panel, mounted above a suspended ceiling, is to be installed. Before the work takes place a check needs to made to confirm the ceiling has:

   a. Perforated accoustic tiles

   b. A suitable light refrative index

   c. An H section grid support system

   d. Sufficient clearance from the structure

4   Before undertaking maintenance work in an office scuff marks and paint damage is identified in the access corridor. This should be notified to the client and:

   a. Covered over during the work

   b. Documented and photographed

   c. Made good on completion of the work

   d. Repaired and reinstated before work begins

5   The protection of any soft furnishings within a work area is best achieved using:

   a. Dust sheets

   b. Corrugated plastic sheet

   c. Plywood

   d. Cardboard

3 Client requirements

RECAP

Before you start work on this chapter, complete the exercise below to ensure that you remember what you learned earlier.

Not _____ that everything is in place for the work to go ahead can result in _____ and additional _____ .

Before work _____ we must identify where we will need to _____ activities with other _____ as failure to do so will generally lead to _____ .

It is important that the maintenance _____ , manufacturer's maintenance _____ and the maintenance _____ are consulted when producing a work _____ .

During the work the local electrical supply is normally _____ and we may need to use _____ power tools and _____ lighting.

Where power is available the use of a _____ tapped _____ low voltage supply system is the _____ option.

Steps and ladders are _____ permissible for work lasting a _____ duration.

It is important to confirm that the test _____ to be used are suitable, are _____ , are free from _____ and _____ correctly.

The lifting equipment must be _____ for use within the work environment and can _____ lift the load.

Access _____ need to be clearly identified and _____ and warning _____ placed to advise _____ of the work.

The work areas, the access and _____ routes and the other areas in the _____ vicinity should be included an existing _____ survey.

Any _____ or defects should be brought to the _____ attention and _____ .

It is important to provide suitable _____ for the fabric, _____ and _____ within the work area.

Dust _____ and _____ floor protectors may be used over the carpets and _____ to protect them during the work.

LEARNING OBJECTIVES

On completion of this chapter you should be able to understand how to determine client requirements for the maintenance of electrical systems. You should be able to:

- Interpret site drawings, plans, maintenance schedules/specifications and the work location

- Interpret appropriate sources of information:

    - Statutory documents

    - Codes of Practice

    - British Standards.

- Evaluate possible proposals to determine how well they meet:

    - Client requirements

    - Site structures and features

    - Industry requirements.

- Identify methods of presenting information to clients to agree and proceed with a plan of work

- State the process and implications that a change in work plans can have in terms of:

    - Health and safety

    - Cost

- Time

- Progress

- Authorization.

● Identify that proposed replacement systems or components comply with industry requirements and where appropriate, give alternative system options which take account of environment and efficiency.

Whilst working through this Chapter you will need to refer to the Institution of Engineering and Technology (IET) Guidance on Electrical Maintenance. This chapter considers how we determine what it is the client requires for their maintenance. The client is often aware that equipment and installations have to be maintained but often is not familiar with what is actually required.

Part 1

Client specifications

Some clients will have a detailed maintenance programme complete with schedules, records and all associated information. However, many clients will produce their own maintenance schedule which is often a simple list of the equipment they want maintained. They may be aware of the location and operation of their equipment, but often they are unaware of what actually needs to be done.

The actual requirements then become the subject of a discussion with the client to ensure that they:

● Comply with their statutory obligations
● Keep the equipment suitably maintained.

Where the client is unsure of the actual requirements we must take care to ensure that we provide the best possible advice and assistance.

Some clients will be aware of the technicalities involved in the work while others will have no knowledge at all. Similarly some will be cooperative while others may be difficult and obstructive.

It is important to establish what it is the client requires and every effort should be made to enable them to achieve it. Clients will generally have at least a rough idea of what they want; some will have quite specific ideas as to what they require. In any event the first thing we need to find out is what the requirements are and how we can ensure that we deliver what they need.

In order to do this we may need to carry out some investigation into the type of equipment and the manufacturer's recommendations. Once these have been obtained we need to discuss these requirements with the client and develop a maintenance schedule. The cost and programme implications for the schedule will need to be determined and agreed with the client.

Once the schedule has been agreed and manufacturer's information confirmed, we need to develop and agree the programme of work.

Figure 3.1 *Agreeing the requirements with the client*

Where the client has a preventative maintenance scheme in operation all these details should be readily available together with the client's records. This makes determining the requirements a lot easier and the programme of works may even be in place. In such cases many of the details may already have been decided but a working programme and contract will still need to be agreed with the client.

It is important to ensure that, irrespective of the size of the maintenance project or the extent of the contractor's involvement in the development, the client is kept informed and updated on progress as the work is carried out.

If your clients appear to have little or no understanding of what the job entails it is important to explain simply and clearly any parts of the work that are not understood. It may be necessary to ask questions of the customers in order to establish the extent of their understanding.

Beyond the technical requirements of the client specification there is also the practical and operational aspect to be considered. The client will have a preference and requirement for the actual work to be undertaken at a time most convenient to them. This will usually include:

- The dates on which work will begin and finish
- The time each day when work can begin to fit in with the client's normal activities
- When work should be finished each day
- When the supplies can be isolated
- Any equipment that has to be operational at the end of each day.

Figure 3.2 *Keeping the client informed as work progresses*

For example, when carrying out maintenance on one floor of an office building the restriction on work activities and noise during the working day is often an issue to be included in the programme. The noise levels, fumes or dust should not cause inconvenience for the occupants of the other floors during their normal working day.

Programme changes need to be discussed with the client

It is important to ensure that the client is advised of changes or developments as they occur. This will ensure the client is aware of any programme or cost implications and make the necessary adjustment to their schedule. This will also minimize any wasted time or materials. This can be a particular issue when equipment requires additional parts above those allowed for in the normal maintenance regime.

Remember

- A courteous approach may result in the customer returning with more work in the future.
- Customers who are not treated with respect will be likely to complain and they may also become aggressive.
- Your employer will expect all staff on-site to be representatives of the company whether dealing with clients, the contractor or subcontractors.

Installation specification

All electrical work will have a specification of some kind. In the case of simple maintenance work it may begin with an informal discussion and verbal agreement between the client and the contractor. This will then become a formal agreement when the quotation for the work is submitted by the contractor. For maintenance work such as the replacement of components the specification may simply be to replace with manufacturer's spares, or the client may accept recognized alternatives providing they match the specification of the original.

At the other end of the scale a full specification is produced, detailing almost everything about the maintenance. The specification may include the details of all the equipment and spares to be used, the circuit arrangements, the type of wiring system and the location of all the equipment.

The specification may also contain details of the programme for the work with start and finish dates together with specific dates for some work activities.

Specification drawings

Figure 3.3 *Drawings and diagrams*

As well as the technical requirements for the maintenance activities and materials the specification may include drawings identifying the location of the equipment and isolation and control devices within the installation. These drawings will be used to determine the requirements for quotation and the maintenance of the equipment.

The details of any control circuits may be included in the specification. Circuit and wiring diagrams provided by the manufacturer and installer for the control of equipment such as motors, lighting and standby generators will be required. These will normally be included in the specification or maintenance record documents.

The specification drawings will generally include layout drawings showing the positioning of equipment and block diagrams showing the sequence of control and distribution equipment.

Drawings and diagrams

The drawings we are most likely to refer to during the maintenance work include:

- Block diagrams
- Circuit diagrams
- Wiring diagrams
- Layout drawings.

Block diagrams

These indicate the sequence of components or equipment. Each item is represented by a labelled block.

Layout diagram

A layout diagram shows the layout of the equipment and the routes of the cables forming the electrical installation. These are produced as part of the installation contract as the 'as fitted' drawings which show how the electrical installation is constructed.

The drawing indicates a scale and the symbols used identify the items of equipment which are installed. This information can then be used to determine the location of equipment within the installation and the circuit routes for control systems.

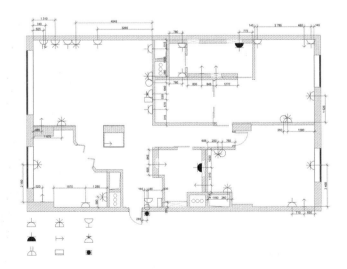

Figure 3.4 *Layout drawing*

Circuit diagrams

Circuit diagrams are used to show how the components of a circuit are connected together. The symbols represent pieces of equipment or apparatus and the diagram will show how the circuit works.

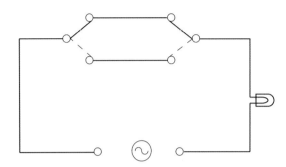

Figure 3.5 *Simple two way switching circuit diagram*

Wiring diagrams

Wiring diagrams indicate the locations of the components in relation to one another and cable connections, and are more detailed than circuit diagrams.

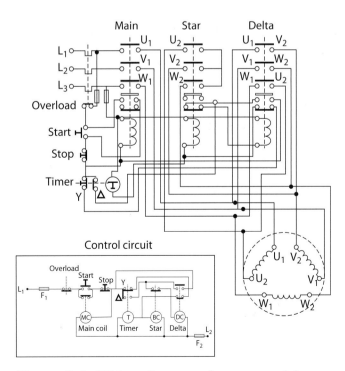

Figure 3.6 *Wiring diagram of auto star delta starter*

Drawing symbols

Our everyday work requires us to use information presented in the form of drawings and diagrams. This may be as a plan or layout to identify the location of components or fittings and accessories, or as diagrams to show the way in which a circuit is connected or how it functions. In addition to these we will also find diagrams used to give the sequence of equipment and controls.

To enable us to refer to these drawings with ease, a system of symbols that can be readily understood and easily interpreted is used. We must first establish what these symbols are and for which type of drawings and diagrams they are used.

BS EN 60617 was the harmonized European Standard which contained standardized symbols. However in 2002 IEC launched an 'on-line' database format for the symbol library, available on subscription from the IEC website. In 2004 CENELEC decided formally to adopt the IEC database and to cease publication of EN 60617 in 'paper' form and to withdraw the then-existing standards. Consequently the British Standard versions have also been withdrawn. There are still copies of the withdrawn standard available for reference but these will not be updated.

The use of standard symbols makes referencing between different drawings much easier. In the maintenance industry, for example, we will often be using drawings produced from several sources. If each of these uses their own set of symbols we will be constantly referring to the key to identify each item which may be time-consuming and irritating.

If all drawings use the same symbols then it means that:

● We will become familiar with the everyday symbols and no referencing will be needed for the majority of work

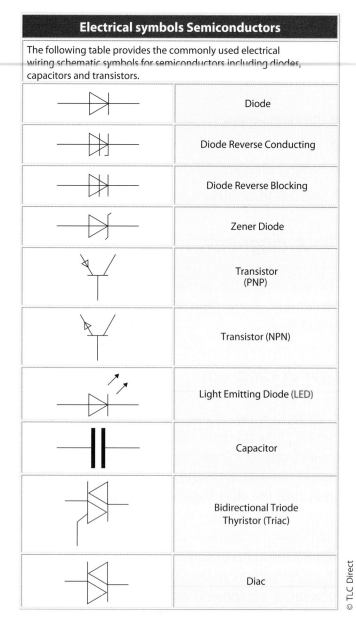

Figure 3.7 *Standard semi-conductor symbols*

● Cross-referencing will be made easier as each drawing will be using the same symbol for each item
● A key will only need to be drawn out for special symbols for unusual items, instead of every item used
● Information will be taken from the drawing quicker and with a lot less stress.

 **Try this**

1   Using the information available for drawing symbols, either as hard copy or on-line, draw the standard symbols for the following:

   a   a cell

   b   a fuse

   c   a resistor

   d   a switch

   e   a variable resistor.

2   Which type of drawing or diagram shows:

   a   simply the sequence in which items are connected?

   b   the electrical operation of a particular circuit or system?

   c   the actual connections involved and the relative component locations?

   d   the position of components and equipment and will be drawn to scale.

Part 2 Sources of information

Before we undertake the maintenance of equipment or wiring systems there are a number of sources of information which we must be aware of and may refer to.

Statutory documents

The principal statutory documents relative to the maintenance of electrotechnical systems are:

- Management of Health & Safety at Work Regulations
- The Electricity at Work Regulations
- Manual Handling Operations Regulations
- Control of Substances Hazardous to Health (COSHH) Regulations
- Workplace (Health and Safety and Welfare) Regulations
- Personal Protective Equipment at Work Regulations (PPE)
- Provision and Use of Work Equipment Regulations (PUWER)
- Work at Height Regulations
- Display Screen Equipment at Work Regulations
- Control of Asbestos at Work Regulations.

We need to be aware of the requirements of these statutory documents where they relate to the electrical maintenance activities.

Codes of practice

A Code of Practice (CoP) comprises a set of written rules which explain how people working in a particular industry should carry out their work. These are common in other areas of industry but in the electrical maintenance industry the CoP for the In-Service Inspection and Testing of Electrical Equipment is the only one regularly used. Much of the other information on the requirements is given in BS 7671 and various guidance notes.

Guidance material

There are a large number of guidance documents available relative to electrical installations. Those commonly referred to include:

BS 7671 Requirements for Electrical Installations

IET Guidance Note 1, Selection and Erection

IET Guidance Note 3, Inspection and Testing

IET On-Site Guide

HSE Guidance Note GS38, Test Equipment Used by Electricians.

Guidance material is produced by recognized industry bodies such as the IET and the HSE and, whilst the guidance notes are not statutory they may be cited in a court of law where a prosecution is brought under statutory legislation.

Electrical **excellence**

Figure 3.8 *IET Guidance material*

British Standards

We have already identified BS 7671, Requirements for Electrical Installations, as one of the guidance documents we may need to refer to. It provides information to enable an electrical installation to be designed, constructed and inspected and tested safely and to be safe for the users of the installation where compliance is achieved. Some of the information contained in BS 7671 is relevant to the maintenance of electrotechnical systems, such as acceptable test results.

There are a number of other British Standards which are relevant to electrical installations. BS 7671, Appendix 1, contains a list of all the British Standards which are referred to in connection with the electrical installation. Some British Standards relate to the manufacture and construction of electrical equipment, such as BS 88-6 for fuses, others relate to installation and performance criteria for electrical systems such as BS 5266 for emergency lighting and BS 5839 for fire alarms.

Figure 3.9 *British Standards Institute*

We will also refer to BS EN Standards; the EN indicates they are a Euro Norm, which means the standard has been aligned with the European requirements.

Task

BS 7671 341-01 states that 'An assessment shall be made of the frequency and quality of maintenance that the installation can reasonably be expected to receive during its intended life.'

Note the three requirements identified there which are to be applied to ensure the installation may be suitably maintained.

1 _____

_____

2 _____

3 _____

Part 3 Presenting the information to the client

Before we can provide the client with a proposal for maintenance we must make sure that the proposal is suitable. When producing a maintenance proposal for the client's installation and equipment it is important to ensure that the proposal:

● Meets all the client's requirements
● Is suitable for the particular site
● Complies with the industry requirements.

Client

In Part 1 of this chapter we identified that the client is not always aware of the requirements for the maintenance of their equipment. Where this is the case the contractor will need to prepare a suitable programme of maintenance for the client. The client will have some expectations even if they have no understanding of the technical and statutory implications.

Following discussions with the client it is important to ensure that the client's requirements and expectations are met.

Site

The proposal must be suitable for the site and this includes the structure and features. The procedures must take account of the structure of the building and allow the work to progress with as little inconvenience to the users as possible. It must also take account of the way in which the building is constructed and how it is laid out. Equipment to be used must be suitable and able to operate within the building without damage to fixtures and fittings. Manoeuvring materials and equipment into position and accessing the work location need to be suitable. Whilst a mobile elevated work platform (MEWP) may be an ideal access method, if this closes off an access or exit route then a fixed scaffold may be required.

Industry

There are many industry requirements to be considered in connection with maintenance and those which apply will depend on the work involved.

Note
Guidance on some of the requirements for maintenance can be found in the IET Guidance, Electrical Maintenance.

We will consider some of the more common electrical maintenance requirements here.

Fixed wiring

BS 7671 identifies the requirement for electrical installations to be periodically inspected and tested and this aligns with the statutory requirements of EWR, PUWER and the other documents mentioned earlier in this study book. The information given in BS 7671 Part 6 and the accompanying IET Guidance Note 3, Inspection and testing, provide further information on the requirements for periodic inspection of fixed wiring. During the periodic inspection any deviations from BS 7671 should be identified allowing them to be actioned and this process forms part of the maintenance of the electrical installation.

Figure 3.10 *Fixed wiring needs to be maintained*

The regular planned preventative maintenance of the electrical installation, with the results of the inspection and tests recorded, will provide the same level of maintenance. For installations where such a regime is in place a periodic inspection may not be required.

Lighting

Lighting maintenance is required as a result of the deterioration of lamps over time which results in a reduced lumen output and inevitable failure. Luminaires and lamps require cleaning and lamp replacement which is determined by the type of lamp and the time it is in operation. The IET Guidance, Electrical Maintenance, includes a table indicating the luminaire maintenance factors, maintenance categories and maintenance factors. This information is used to determine the required maintenance for lighting and should be considered when planning maintenance.

Figure 3.11 *Lighting needs to be maintained*

Electrical Equipment

The CoP for the In-service Inspection and Testing of Electrical Equipment, published by the IET provides guidance on the maintenance of electrical equipment provided in the workplace. This includes electrical equipment provided for use in commercial, industrial and that used by members of the public. The CoP provides guidance on the frequency and inspection and testing requirements for types of equipment categorized as:

- Stationary
- Information Technology equipment
- Movable
- Portable
- Handheld

Figure 3.12 *Electrical equipment*

Task

Before continuing with this chapter, refer to the IET Guidance on Electrical Maintenance and familiarize yourself with the requirements for the maintenance of:

- Electrical installations
- Electrical equipment.

Emergency lighting

The majority of emergency lighting is to provide a suitable level of illumination for the vacation of the premises. It does not provide levels of lighting to allow work to continue as normal, simply to evacuate the building. This is primarily in the event of fire when the general lighting may fail.

In some locations, such as hospitals, there may be a requirement for standby lighting systems which are intended to allow activities in certain areas to continue as normal in the event of a power failure. We shall consider the requirements for emergency lighting here and this may be provided in one of two ways:

- Self-contained luminaires which contain their own batteries and a charger and rectifier within the luminaire and this is supplied from the local lighting circuit
- Centrally controlled where an alternative power supply, generally batteries, is located at a particular location and special circuits are installed to supply the emergency lighting luminaires.

The self-contained system is generally the most common.

Figure 3.13 *Central battery room*

There is a requirement for emergency lighting in a variety of premises and may be categorized as:

- Maintained (M); the luminaires are lit all the time
- Non-maintained (NM); the luminaires are only lit when the general lighting fails.

Both are then further detailed by the period of time for which they can operate, on the batteries, without the main supply being available. This is given as a number so for example an M/2 system would be a maintained system with a 2-hour duration and a NM/1 system would be a non-maintained system with a 1-hour duration.

Figure 3.14 *Emergency lighting*

The maintenance of the system is required and records are to be kept by the client of all the maintenance activities, and these are required:

- Daily (mainly for centrally controlled systems)
- Monthly
- Six-monthly
- Yearly (for self-contained luminaires)
- Three-yearly.

BS 5266 contains the requirements for emergency lighting systems and guidance and examples of records are contained in the IET Electrical Maintenance.

Fire alarms

Fire alarms are required in most buildings and all buildings used by the public. There are two standards which identify the requirements for fire alarm systems:

- BS 5839-1 for Buildings
- BS 5839-6 for Dwellings.

As with the emergency lighting there is a requirement for regular maintenance on fire alarm systems and the system may be installed to provide:

- Protection of life
- Protection of property (may include protection of the environment and business loss).

The category and type of fire alarm system determines the protection and operation of the system with category L denoting life protection and category P denoting property protection.

For example a category:

- M is a manual system with no automatic fire detectors
- L1 is a life protection system installed throughout all areas of the building to give the earliest possible warning and longest evacuation time
- P1 is a property protection system installed throughout the building to give early warning of fire and minimize the time between the starting of the fire and arrival of the fire services.

Figure 3.15 *Fire alarm manual call point*

The user of the installation, generally our client, is required to maintain a log of the fire alarm testing and maintenance. The tests fall into four main categories:

Weekly: for the majority of systems to ensure the fire alarm system is functioning correctly.

Monthly: for larger systems with standby generators or 'vented batteries'

Quarterly: for larger systems with standby generators or 'vented batteries'

Periodic inspection and test: determined by a risk assessment.

Switchgear

Commercial and industrial switchgear, both low voltage (LV) and high voltage (HV) needs to be regularly maintained. The Health and Safety Executive (HSE) produce guidance in the form of HSG 230, 'Keeping Electrical Switchgear Safe'.

Note

HSG 230 is available as a free download from the HSE website.

There are many types of switchgear and each manufacturer will have operation and maintenance details for their particular switchgear.

The main types are:

- Oil
- Gas
- Air
- Vacuum.

and these are the mediums used within the switchgear to extinguish the arc produced during their operation.

In general terms the maintenance schedule should consist of a minimum of:

- Annual inspection
- Five-yearly examination
- Overhaul at maximum of 15 years.

Figure 3.16 Main switchgear

Task

Before continuing with this chapter refer to the IET Guidance on Electrical Maintenance and familiarize yourself with the requirements for the maintenance of:

- Emergency lighting
- Fire alarms
- Switchgear.

Part 4 Informing the client

Most companies have their own company process and procedures for presenting information to the client and any other parties involved in the maintenance work.

This will generally include a document register in which the issue and receipt of information is recorded. A client file is normally generated which will contain all the relevant information and correspondence to and from the client.

Remember

The client is the person or organization which employs your company to carry out the work.

When providing information it is important to remember that some clients may be aware of the technical issues involved while others may have no knowledge at all. It is important to make sure that in your dealings with the client and other relevant parties we ensure that:

- All the information provided is accurate and complete. The client will be reliant on the information you provide so that they can agree and allow you to proceed with the plan of work
- Copies of the information should always be kept, as discussed earlier, to ensure that there is a record of the information provided and to enable help and advice to be provided to the client in the future. It is important that the information related to the maintenance is retained for contractual and legislative reasons. The outcome of any contractual dispute which may arise could be dependent upon the records kept by both parties.

We need to make sure that the information is provided in the most suitable and user friendly format and is:

- Accurate
- Clear and concise
- Provided courteously and professionally

by whatever method is used to provide it.

The use of IT allows a great deal of information to be provided electronically. This has a number of advantages including that it:

- Is environmentally friendly (less paper)
- Requires reduced storage facilities
- Is easier to transport and retrieve.

So wherever it is possible and practical this method should be considered.

That is not to imply that paper information is not required. Layout drawings, plans and similar information will still be required in paper format on-site. Many day work instructions and

variation orders are provided in paper format, generally because the use of IT equipment to provide such information on-site is not practical or convenient.

Remember

Methods of communicating with your client include:

- **Face to face**
- **Telephone**
- **Facsimile**
- **Mail**
- **Courier**
- **Email**
- **Data format (discs, recordable medium, etc).**

Figure 3.17 *Information on paper is still convenient*

Record information relating to the contract, site activities, method statements and so on are well suited to electronic delivery and storage. Recipients are able to review and if necessary print off hard copies for themselves if they are required.

A suitable covering letter for material and written information delivered, by post or email, is important. In much the same way as a drawing issue sheet details exactly what is provided the covering information outlines the content. It also helps to ensure that the information is given to the appropriate person or persons.

When providing the operation and setting of controls information face to face delivery will often be useful for the client particularly when accompanied by a practical demonstration of the process. Whenever information is imparted by face to face delivery it is essential that the client is not made to feel intimidated or belittled.

A diplomatic approach in the early stages is important to determine the level of understanding without causing offence. Once the client's level of understanding is established, we must take care to keep the client informed in terms which are appropriate to their level of technical understanding. So the terminology used will be different when communicating with an electrical consultant compared with a layperson.

Figure 3.18 *Good communication with the client is important*

The details given to the client should identify:

- What work is required
- The extent of the work
- Any spares or replacements which may be necessary and
- An indication of the time and cost involved.

This will allow the client to discuss and agree a programme of work and issue a contract.

Having provided the appropriate information, it is also important to make sure that it is understood. Allowing the client to ask any questions regarding the information they have been given may not be sufficient, particularly when dealing with customers who are non-technical. It is often a good idea to offer some brief verbal explanation outlining what has been provided and, where the customer appears to be uncertain, offer some further help or a demonstration where it is appropriate.

Remember

- It is important to make sure that all your dealings with the customer are carried out in a polite and courteous manner.
- Do not talk down to the customer.
- Do not provide explanations in very technical language unless the customer is familiar with the electrical maintenance industry.

 **Try this**

You have been asked to carry out the maintenance of an electrical installation within a small retail outlet which is under new management. The sales area is over two floors and a basement level contains the staff and storage facilities. To comply with the statutory requirements the premises has a fire alarm over all floors and emergency lighting provided by self-contained luminaires.

1 Describe which method(s) you would use to provide the client with the details of the work and the proposed programme.

_____

_____

_____

2 Explain how you would confirm that all the client's statutory obligations are included in the proposal.

_____

_____

_____

3 List the industry standards that you will need to refer to when preparing the work programme and carrying out the maintenance.

_____

_____

_____

Implications of a change in work plans

Having agreed the extent of the work and a work programme it is important to advise the client as soon as possible of any changes and the implications they will have on the work.

Often work plans have to change because of such things as:

- Client requirements
- Additional works
- Permit to work requirements
- Environmental issues
- Deliveries of spares or equipment
- Availability of personnel.

These changes can in turn affect:

- Health and safety
- Cost
- Time
- Progress.

The health and safety requirements may be affected by changes to the work requirements and environmental conditions. For example work was to be carried out whilst the production machine shop was idle. An order placed with the client means that the maintenance has to be carried out whilst full production is ongoing. Additional protection because of the noise levels and possibility of pollutants in the atmosphere needs to be considered.

Additional work or changes to the work requirements can have a significant effect on the work programme and cost. These are generally discussed with the client in the first instance and followed by written details identifying the additional costs.

Any additional work requirements can also have an effect on the programme that has been agreed. It is important that the programme is monitored and any change in the progress of the work is reported and dealt with as soon as possible as the contract may contain a penalty clause for delay to the completion of the work. This may result in the company responsible for the delay paying a considerable penalty.

Any changes to the level of personnel available for the work or changes in the availability of spares, materials or plant may also have an effect on the programme. For example one member of the maintenance team going off sick will affect the progress of the work.

Replacement systems or components

The type and manufacturer of the spares and equipment required for electrical maintenance work is normally identified in the specification. Only spares produced by the manufacturer of original equipment may be specified and we are obliged to provide genuine spares.

The availability of spares plays an important role in making a decision as to whether we provide new, replacement or refurbished components. Alternatives to the spares and equipment specified may be considered if these have an effect on:

- The timing of the work
- Efficiency
- The environment

and would need to be agreed with the client before any alternatives or changes are made.

The use of alternative 'pattern' spare parts may be considered by the client where they offer a saving in either cost or time for the work to be completed. In some instances it may be that a direct manufacturer's replacement is no longer available. A pattern alternative or even a reconditioned part may be the only alternative in such cases.

The client may be advised that the use of an alternative part could produce an increase in efficiency. For example changing lamps to low energy types when replacing them as part of a maintenance programme can produce energy savings. An alternative would be to consider the use of presence sensors to control lighting which would also produce a saving using lighting more efficiently.

Figure 3.19 *Movement sensor to control lighting*

Environmental savings may also be considered by the client for a number of reasons. Most companies are being actively targeted to reduce their carbon footprint and any measures that can be offered to achieve this may be favourably received. The efficiency examples above would result in reduced cost and lowering the carbon footprint and so may be worthy of consideration.

Many of the environmental benefits can result in a cost saving and better efficiency but some may be considered because the company has a green policy and is in favour of such changes. For example when maintaining the heating system a client may be inclined to consider the use of solar thermal energy as part of their maintenance. This may not be a direct saving after the cost of installation and equipment is taken into

account. However the client may be inclined to install such a system for the environmental benefits it will provide. Of course this can also help to reduce the carbon footprint of the premises.

Note

Further information on renewable energy options is given in the Legislation study book in this series.

Any replacement components or equipment must comply with industry standards. Where an alternative option is considered then these must also comply with the industry standards.

Task

Prepare written information for the client identifying suitable options for the following. (Manufacturer's data either written or on the Internet may be used to help with this task.)

1 Recommend alternative options for the replacement of a component in a valuable machine which is now obsolete and spares from the manufacturer are no longer available.

_____

_____

_____

2 Improving efficiency when replacing diacroic downlighter lamps as part of a maintenance programme.

_____

_____

_____

3 Improving the environmental impact of a gas-fired hot water system for an office building.

_____

_____

_____

Congratulations you have now completed Chapter 3 of this study book. Complete the self assessment questions before continuing to the next chapter.

SELF ASSESSMENT

1 Whilst carrying out maintenance on a motor circuit which includes an auto star/delta starter the electrician may refer to the:

   a. Wiring diagram

   b. Circuit diagram

   c. Block diagram

   d. Layout diagram

2 ⎯⎯▷|⎯⎯ The symbol represents a:

   a. Diode

   b. Reverse conducting diode

   c. Reverse blocking diode

   d. Zener diode

3 'A set of written rules which explain how people working in a particular industry should carry out their work' describes a:

   a. British Standard

   b. Statutory Regulation

   c. Code of Practice

   d. Guidance Note

4 An emergency lighting system which is non-maintained and has a 1-hour performance is categorized as:

   a. NM/1

   b. NM/2

   c. M/1

   d. M/2

5 The client has asked that the maintenance in one area of the building is to be done out of normal working hours. The client will need to be advised of the effect this will have on the cost of the work and the:

   a. Time to complete the work

   b. Availability of spares

   c. Provision of labour

   d. Completion of records

Work requirements for maintenance

4

RECAP

Before you start work on this chapter, complete the exercise below to ensure that you remember what you learned earlier.

A discussion with the _____ is required to establish their _____ and ensure their compliance with the _____ requirements and keep their _____ suitably _____.

It is important to ensure that the client is kept _____ on progress as the _____ is carried out.

The practical and operational considerations include: when the supplies can be _____ and any equipment that must be _____ at the end of _____ day.

It is important that the client is _____ of any programme or _____ implications as soon as possible to allow them to _____ their _____ and _____.

A circuit diagram shows how the _____ of a circuit are _____ together and wiring diagrams indicate the _____ of the components in _____ to one another and cable _____.

Any maintenance _____ must meet the client's _____ be suitable for the _____ worksite and comply with the _____ standards.

Industry standards need to be consulted for the _____ wiring, _____, electrical _____, fire alarms, _____ lighting and _____.

Any information must be in a _____ and user _____ format and be _____, clear and _____.

The details given to the client should identify the _____ which is required, the _____ of the _____, any _____ or replacements which may be necessary.

Work plans may have to change due to client _____, additional _____, _____ issues, availability of _____ which can affect _____ and safety, costs and _____ .

LEARNING OBJECTIVES

On completion of this chapter you should be able to:

- State the characteristics of different types of electrical maintenance activities, including:

  - Planned preventative

  - Breakdown

  - Monitored

  - Non-routine maintenance.

- Specify the importance of:

  - Agreeing start dates, finish dates and timings

  - Procedures for agreeing variations to the maintenance specification or schedule.

- Define the specific range of job information that is required for maintenance work including:

  - Statutory documents

  - Codes of Practice

  - British Standards

  - Manufacturer's specifications

  - Legal requirements for maintenance:

    - Common Law requirements

    - Specific legal requirements (plant/equipment)

    - Implied legal requirements.

- State how specific job information can be used to help develop work proposals

- Specify the replacement/refitting requirements for components within maintained electrical systems and equipment.

- State appropriate methods for determining the size and specification/type of components to be used when maintaining electrical systems and equipment

- Interpret drawings and maintenance schedules/specifications to calculate resources required to complete electrical maintenance work with regards to:

  - Materials for plant, equipment and components for use within maintenance programmes

  - Tools and equipment.

- Identify the implications that different working conditions could have on equipment and components in an electrical installation.

Whilst working through this Chapter you will need to refer to the Institution of Engineering and Technology (IET) Guidance on Electrical Maintenance. This chapter considers how we determine what is required for the maintenance work to be undertaken. This includes the type of work involved, the necessary information, tools and materials together with the work programme.

Part 1 Types of maintenance

Maintenance may be carried out in one of four ways.

1   Preventative maintenance: where maintenance is carried out in accordance with a maintenance schedule it is planned and carried out at specified intervals and in a particular way.
2   Breakdown maintenance: is simply carried out when there is a breakdown, i.e. when equipment or systems cease to work.

3   Condition monitored maintenance: where the condition of the system or equipment is monitored by measurement and alarms are used to indicate the need for shutdown and maintenance. These alarms are used when values go above or below predetermined parameters and indicate the need for maintenance to be carried out.
4   Non-routine maintenance: where maintenance is required which has not been scheduled or demanded by failure.

Preventative maintenance

Preventative maintenance is a schedule of planned maintenance tasks aimed to prevent the failure of equipment or systems. Examples of preventative maintenance are the changing of street lighting lamps on a motorway or in a lighthouse. Whilst these lamps may not have failed, the maintenance periods are set to times when their efficacy and life expectancy are reduced to an unacceptable level.

Figure 4.1 *Motorway lighting*

Figure 4.2 *Lighthouse*

Breakdown maintenance

This is when equipment and systems are replaced when they no longer work. When a breakdown does not result in a dangerous situation then this is the simplest type of maintenance. It is typically used in such situations as replacing failed lamps in a domestic house.

Figure 4.3 *Replacing a lamp*

Condition monitored maintenance

Monitored maintenance is where certain parameters are measured and when they reach preset levels that is the time for the equipment to be shut down and maintenance to be carried out. This includes monitoring of temperature, vibration and pressure with sensors to monitor the condition of plant and equipment.

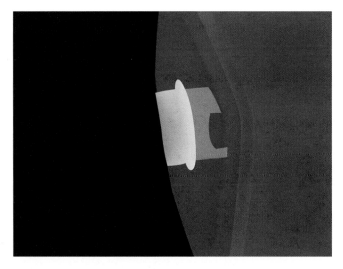

Figure 4.4 *Thermal image showing a bearing overheating*

Non-routine maintenance

Non-routine maintenance is where maintenance is required which has not been scheduled or demanded by failure. One example of a non-routine maintenance task is one that may have to be carried out when a manufacturer has issued a service bulletin to correct a problem with a product. The problem could indicate that because of design or manufacturing problems the device produced may be potentially defective. It may also be that a product has to be recalled or withdrawn from the market and this would require non-routine maintenance.

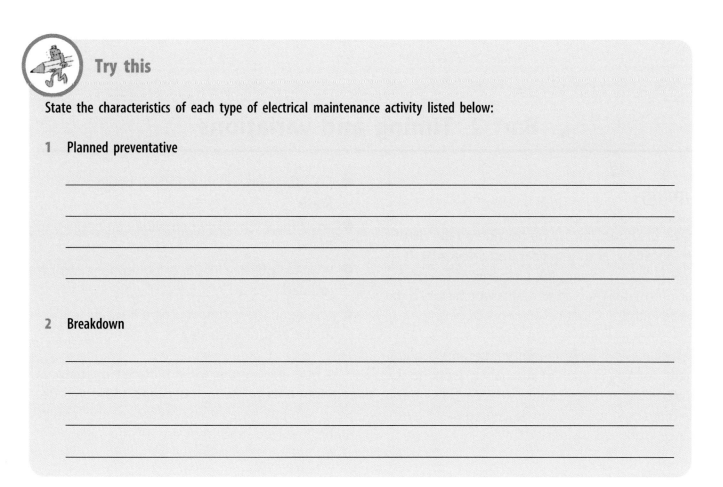

Try this

State the characteristics of each type of electrical maintenance activity listed below:

1 **Planned preventative**

_____

_____

_____

_____

2 **Breakdown**

_____

_____

_____

_____

3 Monitored

_____

_____

_____

_____

4 Non-routine maintenance

_____

_____

_____

_____

Part 2 Timing and variations

Timing

In the previous chapter we considered the client's specification and practical requirements. It is worthwhile reviewing the elements of the practical information and agreements which need to be made with the client before work begins.

The client will have a preference and requirement for the actual maintenance work to be undertaken at times causing the least disruption to them and their business. For most maintenance work this will include:

- The dates on which work will begin and finish
- The time each day when work can begin and finish
- When supplies, circuits or equipment can be isolated
- The equipment that must be back in service at the end of each day.

It is not uncommon for additional operational requirements to be included and these will depend on the type of maintenance, the equipment involved and the use of the premises.

For example, when carrying out work in a pottery workshop the restriction on work activities relating to the firing of the kilns is often quite specific. Once the firing has begun it must follow its full course or the content will be ruined. The progress of the firing should allow production to proceed with the minimum of disruption.

It is important to discuss any programme changes with the client and to obtain their agreement. It may be that the client approaches us with a proposed change due to a change in their workload. In such instances we must consider the programme implications and advise the client of any knock-on effects. These changes may result in changes to the finish date or involve out of hours working or an increase in the number of operatives to achieve the desired end date.

Remember

The effect of any changes to the agreed schedule must be detailed for the client to enable an agreement to be reached on the best way to accommodate the changes.

During the work process there are many events that may require changes in the scheduled programme of work. Some of the main events which often lead to reprogramming maintenance activities include:

- Delays due to weather, particularly with outdoor activities, although in severe winter conditions access to the worksite or the travel of operatives may not be possible
- Unforeseen circumstances such as:

  – The failure of plant or equipment
  – The sickness or injury of operatives
  – Non-delivery of spares, plant or equipment
  – Unexpected component failures in equipment being maintained.

Figure 4.5 *Changes in programme may have serious effects for the client*

It is essential to monitor the programme once the work begins and that any changes are reported and dealt promptly.

Remember

What appears to be only a minor delay to our work could have major implications for the client's business.

Task

A specialist supplier has informed you that the parts required for the maintenance of one particular specialized shot blasting machine will be 3 days late in arriving from the manufacturer. This machine was to be maintained together with the five other shot blasting machines on day 3 of the maintenance schedule. The parts were scheduled for delivery on day 2 of the work programme and the whole maintenance programme runs for a total period of 8 days.

Detail the information that you would provide to the client with regarding this delay and offering the client an alternative option to ensure the programme finishes on time.

_____

_____

_____

_____

_____

_____

_____

_____

Changes to the specification or schedule

Changes to the specification or the work schedule may be generated by the client or by the contractor carrying out the maintenance (us). Any changes will require documentation to confirm the change has been agreed and this is normally done by the issue of a variation order.

Variation orders

Variation orders are issued when any changes to the original specification or schedule are required. On larger contracts a client may have their own electrical engineer or an electrical consultant who would issue the variation order on the client's behalf.

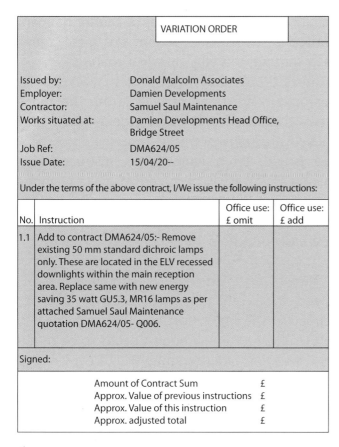

Figure 4.6 *A typical Variation Order*

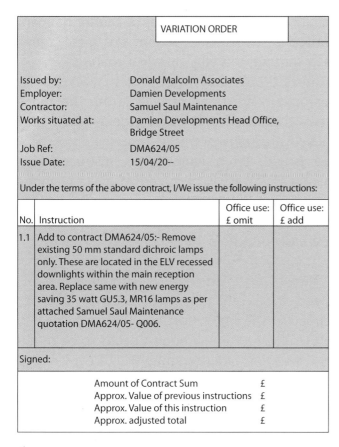

Figure 4.7 *Typical variation request form*

such as a change to an item of equipment or supplier. The change may offer benefits such as a shorter delivery time or a reduced cost without loss of quality and a request for a variation will need to be made to the client. The request for a variation order will need to include a justification for the change identifying the advantages, and cost implications and effect on the programme.

If the proposal is accepted a variation order will be issued to confirm the change to the specification.

Most companies have their own forms for variation requests and variation orders but they all contain similar information. The process of dealing with variations may also vary from company to company.

Confirmation of receipt of variation

We must notify the person issuing the variation order of its receipt and whether it is accepted. There may be occasions where a variation will be questioned due to the need for further information, clarification or for practical reasons.

On smaller projects the client will often approach you directly with a verbal request for any changes to the agreed work. When this happens you should discuss the changes with the client to determine exactly what is required. You should then confirm, with the client, that the information you have recorded is correct and advise of any possible effect on the maintenance programme as a result.

You may need to forward the information to your company office, together with an estimate of the time and any materials involved. This will allow them to prepare a quotation for the change and advise on any effect on the programme.

There may be occasions where you need to request a change to the agreed specification

Most companies have a standard form to do this job and a typical example is shown in Figure 4.8.

Upon receipt of a variation order on-site we must ensure that we record the instruction number, the date it was received and a brief description of the content. We must then investigate what is required and advise the client, within the stipulated time period, of any financial or programme implications.

Samuel Saul Maintenance

To: Contract No.
                  Client Ref:
                  Date:

Dear Sirs, Re: CONFIRMATION OF RECEIPT OF INSTRUCTION   CONTRACT: We acknowledge receipt from your representative......................... of instructions to vary our works as described below: We confirm that in accordance with the terms and conditions of the contract between *(Co. name.............................)* and *(Client name...........................)* the above identified/described instruction(s) which have been received by *(Co. name....................)* will be valued in accordance with the terms and conditions of the contract. The above instruction may result in a variation to our contract price and programme. Signature of Client's rep. Issued by:

Figure 4.8 *Confirmation of receipt of instruction*

However there are some occasions, particularly during maintenance activities, when this variation procedure does not or cannot happen, particularly where instructions are given other than through the official formal channels.

Instructions are often received from the client when the variation is of an urgent nature and such instructions normally take the form of a site issued, often handwritten, instruction. In many cases we will be asked to make changes verbally. We must notify the receipt of verbal instructions and confirm exactly what it is we have been asked to do.

A site issued instruction will normally be followed by a variation order for the work described in the site instruction. Where this is not forthcoming a formal request needs to be made for a variation order together with a copy of the site instruction.

Confirmation of the receipt of variation instructions is to be sent irrespective of how the instruction is received or from whom. Many contracts state that if notification of receipt is not refuted, usually within 14 days, then it is deemed that the instruction has been accepted and we are entitled to be paid for the work involved.

Remember

A variation order may not carry any cost or programme implication but without it we will be obliged to install to the original requirements. If the variation order is not issued then we will not be paid for any uninstructed variations.

Task

The 20 recessed LED replacement luminaires specified as part of a contract are supplied with heat resistant tails but without terminal boxes. The terminal boxes must contain the connections between the luminaire and supply cable and provide clamps to secure the cables.

Using manufacturers' information select a suitable terminal box for this task and complete the Variation Request Form in Figure 4.9 for the client for the supply of the boxes selected.

Variation Request			
Samuel Saul Maintenance			
Contract Ref.	PT573/02	Contract address	Solomon Engineering, Unit 27 Canal Street, Wendingham, WD1 9AT
Date:			
Drawing Ref.	N/A		
Request			
Costing	Delete item: none	£0.0	Replace Item:	£
Programme			

Figure 4.9 *Variation order*

Part 3 Job information

Let's begin by recapping on the regulatory information relevant to electrical maintenance. Regulatory requirements are either in the form of statutory or non-statutory publications. Statutory requirements place responsibilities on us which are enforceable by law. Non-statutory publications are basically guidance and whilst they do not have a legal status they may be quoted in a court of law in the event of a prosecution under statutory legislation.

The main statutory publications which affect electrical maintenance are:

- The Health and Safety at Work (etc.) Act (HSWA)
- The Electricity at Work Regulations (EWR).

The HSWA applies to all work activities and under this statute there are numerous statutory regulations relating to all work activities. It is under this statute the EWR is placed and this has a direct and specific relevance to our electrical maintenance work.

The requirements of the statutory documents are quite simple in essence – an electrical system must be maintained so that it is safe for use and will not give rise to danger.

Remember

An electrical system is a source of electrical energy, a current-using device and the connections between the two.

BS 7671, Requirements for Electrical Installations, the IET Wiring Regulations published by the Institution of Engineering and Technology is the principal guidance for electrical installations. Whilst this is not a statutory document it is accepted as standard practice for electrical installation work and may be cited in a court of law.

BS 7671 contains information on the requirements relating to the electrical installation and any maintenance of the electrical installations must be carried out in accordance with these requirements.

There is a relationship between the requirements of the Electricity at Work Regulations and BS 7671 which is shown in Table 4.1.

Table 4.1 *Comparison of BS 7671 and EWR*

Electricity at Work Regulations	BS 7671
Systems, work activities and protective equipment	Parts 1 and 3
Strength and capability of electrical equipment	Parts 1 and 5
Adverse or hazardous environments	Parts 1 and 7
Insulation placing and protection of conductors	Parts 1 and 5
Earthing or other suitable precautions	Parts 1, 4 and 5
Integrity of reference conductors	Parts 4 and 5
Connections	Parts 1, 5 and 7
Means of protection from excess current	Parts 1 and 4
Means for cutting off the supply and for isolation	Parts 1, 4 and 5
Precautions for work on equipment made dead	Parts 1, 4 and 5
Work on or near live conductors. Working space, access and lighting	Parts 1, 4, 5 and 7
Persons to be competent and prevent danger and injury	Parts 1 and 2

We can see from Table 4.1 that many of the requirements of EWR are addressed in BS 7671 but there is no reference for working on or near live conductors. This is because:

- EWR permits working on live conductors providing it is reasonable in all circumstances for the work to be carried out and that appropriate precautions are taken to prevent injury.
- So whilst live testing during maintenance may be justifiable, there is no justification for any other work to be carried out live.

In addition to BS 7671 the IET produces other guidance material in the form of Guidance Notes designed to provide further guidance on the requirements of BS 7671. The IET Guidance on Electrical Maintenance is a valuable reference for the requirements of electrical maintenance.

There are a number of other British Standards and Codes of Practice (COP) which may apply to the maintenance activities and these include:

- BS 67: Specification for ceiling roses
- BS 95: Electrical earthing. Clamps for earthing and bonding. Specification
- BS 1363: 13A plugs, socket-outlets, connection units and adaptors
- BS 5733: Specification for general requirements for electrical accessories
- BS 8488: Prefabricated wiring systems intended for permanent connection in fixed installations
- BS EN 50107: Signs and luminous-discharge-tube installations operating from a no-load rated output voltage exceeding 1 kV but not exceeding 10 kV
- BS EN 50174: Information technology – Cabling installation
- BS EN 50281: Electrical apparatus for use in the presence of combustible dust
- BS EN 60079: Electrical apparatus for explosive gas atmospheres
- BS EN 60309: Plugs, socket-outlets and couplers for industrial purposes
- BS EN 60598: Luminaires
- BS EN 60669: Switches for household and similar fixed electrical installations
- BS EN 60947: Low-voltage switchgear and control gear
- BS 5730: Monitoring and maintenance guide for mineral insulating oils in electrical equipment
- BS 5266: Emergency lighting
- BS 5839: Fire alarms.

This list is not exhaustive and there are many BS and BS EN Standards which have information and requirements related to the equipment and systems to be maintained. Some of these are quite detailed whilst some are quite general.

There are also CENELEC (the European electrical standards body) harmonized documents and the requirements from many of these are incorporated in BS7671.

Approved CoP are non-statutory but may be cited in a court of law where such a CoP is relevant to a matter of prosecution. The most common CoP used in connection with electrical equipment is the Code of Practice for In-service Inspection and Testing of Electrical Equipment. This CoP does include requirements for connections, in particular those associated with plugs. The Code of Practice for Electrical Vehicle Charging Equipment Installation is a further example of a CoP relevant to electrical installation and maintenance.

Manufacturers' information is generally provided with electrical equipment and accessories. These instructions often refer to the requirements for connection of the equipment. This may include the right connections for the equipment to function correctly and any special requirements such as torque settings, special encapsulation or sealing procedures and the requirements for enclosures. They will usually contain any maintenance requirements such as the type and frequency of maintenance.

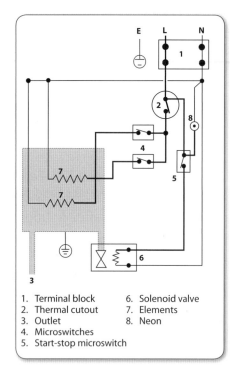
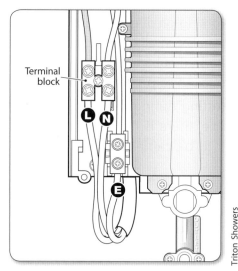

1. Terminal block
2. Thermal cutout
3. Outlet
4. Microswitches
5. Start-stop microswitch
6. Solenoid valve
7. Elements
8. Neon

Triton Showers

Figure 4.10 *Typical manufacturer's instruction for termination*

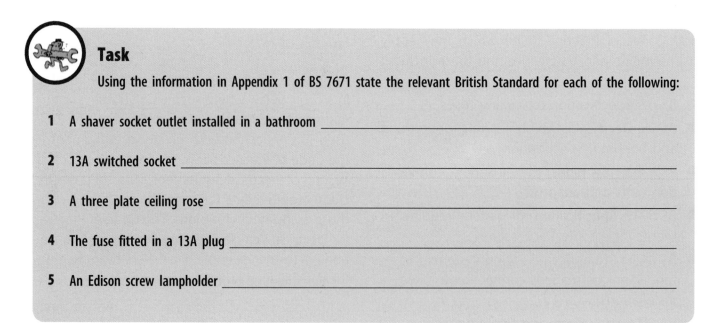

Task

Using the information in Appendix 1 of BS 7671 state the relevant British Standard for each of the following:

1 A shaver socket outlet installed in a bathroom _____

2 13A switched socket _____

3 A three plate ceiling rose _____

4 The fuse fitted in a 13A plug _____

5 An Edison screw lampholder _____

Legal requirements

These are considered in three main categories:

● Common legal requirements

● Specific legal requirements

● Implied legal requirements.

Common law: this implies a general duty of care to other people, property and livestock.

Specific legal requirements: a common example of this is the Health and Safety at Work (etc.) Act 1974 which allows the Secretary of State to make regulations related to health and safety.

The Act also allows the Health and Safety Commission to produce Codes of Practice either directly or with a third party. Under this Act there are many health and safety regulations and some of these include a requirement for maintenance.

The Lifting Operations and Lifting Equipment Regulations 1998 (LOLER) in Regulation 9 places a duty on the employer:

'To ensure that lifting equipment which is exposed to conditions causing deterioration which is liable to result in dangerous situations is thoroughly examined.'

LOLER defines thorough examination as *'a thorough examination by a competent person'*.

Figure 4.11 *Lifting equipment*

In general terms this inspection is required as follows:

● Lifting equipment for lifting persons or an accessory for lifting is carried out at least every 6 months;

● Other lifting equipment at least every 12 months or in either case

● In accordance with an examination scheme and

● When exceptional circumstances occur which are liable to jeopardize the safety of the lifting equipment

● Where appropriate, lifting equipment is inspected by a competent person at suitable intervals between thorough examinations and

● Where it is appropriate this includes such testing by a competent person as is appropriate for the purpose.

The reason for these requirements is given as:

'To ensure that health and safety conditions are maintained and that any deterioration can be detected and remedied in good time.'

Taking these requirements into consideration we can see that there is a statutory obligation to maintain the equipment specifically within the legislation.

Such specific legislation applies to equipment such as lifts, hoists, cranes, cradle systems and electrical equipment in certain premises.

Similarly there is a requirement for working at height and the HSE produce an information sheet, Construction Information Sheet No 47 (rev1) giving recommendations together with a standard inspection report form.

A scaffold inspection report should note any defects and corrective actions taken. Even if the corrective actions are taken promptly they should still be recorded as this helps in the identification of any recurring problems.

Place of work or work equipment	Timing and frequency of checks, inspections and examinations								
	Inspect before work at the start of every shift (see note 1)	Inspect after any event likely to have affected its strength or stability	Inspect after accidental fall of rock, earth or other material	Inspect after installation or assembly in any position (see notes 2 and 3)	Inspect at suitable intervals	Inspect after exceptional circumstances which are liable to jeopardise the safety of work equipment	Inspect at intervals not exceeding 7 days (see note 3)	Check on each occasion before use (REPORT NOT REQUIRED)	LOLER Thorough Examination (if work equipment subject to LOLER) (see note 4)
Excavations which are supported to prevent any person being buried or trapped by an accidental collapse or a fall or dislodegement of material	✓	✓	✓						
Cofferdams and caissons	✓	✓							
The surface and every parapet or permanent rail of every existing place of work at height								✓	
Guard rails, toe boards, barriers and similar collective means of fall protection				✓	✓	✓			
Scaffolds and other working platforms (including tower scaffolds and MEWPs) used for construction work and from which a person could fall more than 2m				✓		✓	✓		✓
All other working platforms				✓	✓	✓			✓
Collective safeguards for arresting falls (eg nets, airbags, soft landing systems)				✓	✓	✓			
Personal fall protection systems (including work positioning, rope access, ork restraint and fall arrest systems)				✓	✓	✓			✓
Ladders and stepladders					✓	✓		✓	

HSE

Figure 4.12 *Inspection table from HSE Construction Information Sheet No 47(rev1)*

Implied legal requirements: these are more common in the legislation, for example Regulation 4(2) of EWR states

'As may be necessary to prevent danger, all systems shall be maintained so as to prevent, so far as is reasonably practicable, such danger.'

This implies that the system, which includes the equipment, is to be maintained so it is safe. To keep the system in a safe condition will normally require maintenance. In turn the maintenance will require inspection, and where appropriate testing, to determine the condition and whether maintenance is required. So the inspection and testing is effectively part of the maintenance activity.

Records of all these activities and the results of inspection, any testing and any repairs are kept to demonstrate that the legal obligations have been carried out. In addition the records will allow the effectiveness of any maintenance to be determined and any necessary changes to frequency or extent of maintenance to be made.

The in-service inspection and testing of electrical equipment is one of the procedures, together with periodic reporting, used to meet the statutory requirements of EWR. The IET publication Code of Practice for In-service Inspection and Testing of Electrical Equipment provides guidance on this process. By reviewing and analyzing the results of the inspection and testing a risk assessment can be undertaken. This can be used to determine the ongoing frequency of the inspection and testing activities for the specific equipment and environmental conditions.

Task

Using the information contained in the Code of Practice for Inspection and Testing list the statutory documents which require equipment to be maintained in a safe working condition.

_____

_____

_____

_____

_____

_____

Developing work proposals

In Chapter 1 of this study book we considered the information required to allow maintenance to be carried out safely and produce a work programme. We now need to consider the specific information that can be used to help develop work proposals.

This information can include such items as drawings, diagrams, maintenance schedules, specifications, data charts, manufacturer's manuals, servicing records and standard maintenance time records. We can best review this process using an example of typical maintenance activity, such as the maintenance of an electric saw in a workshop.

The first item of information we require is the client's specification. For this particular client the circuit supplying the saw is also to be inspected and tested to confirm it meets the current requirements of BS 7671. From the client's specification we can see that the activities can be considered as two main elements:

● The inspection and test of the circuit
● The inspection, testing and maintenance of the saw motor.

The circuit

When maintenance takes place on an electrical circuit or installation then any previous Electrical Installation Condition Reports (EICR) (formerly periodic inspection reports (PIR)) should be available. Where the installation has not been subject to such an inspection before then any certification, such as the Electrical Installation Certificate and any Minor Electrical Installation Works Certificates would be requested.

Where this information is available we can identify the isolation and switching arrangements, the circuit identification, protective devices and so on. Where such information is not available the maintenance work will need to include a survey to establish these details. The information on these documents will also provide evidence of the previous condition of the electrical installation and help to identify any deterioration that may have taken place.

Some of the information should be contained on the circuit charts for the installation and the absence of these will also require some survey and investigation work for the maintenance to be carried out safely.

The motor

The maintenance requirement for equipment is often similar to that of a car maintenance scheme. The interval between certain activities varies and the maintenance manual or schedule identifies what is required at each interval.

To maintain the motor some of the information is essential to allow the work to be undertaken. This information may be part of the specification, such as the actual details of the work to be carried out. Often the information is contained in the manufacturer's information for the machine.

The information contained in the manufacturer's literature which is required for the maintenance will include:

- Technical information such as power requirement, current and voltage rating, settings for controls, overloads and so on
- Wiring diagrams
- The type and frequency of maintenance activities
- Fault finding charts
- Spare parts information and references.

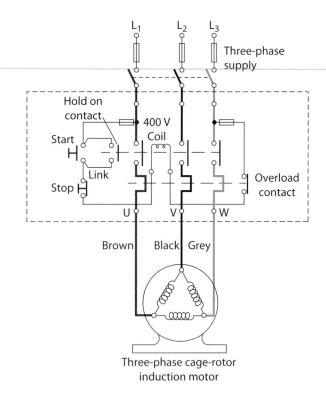

Figure 4.13 *Wiring diagram for three-phase motor and starter*

Having this information available will allow the maintenance to be carried out efficiently and give an indication of the work required. This will allow an estimate of the time required for the work and format a work proposal.

Any existing maintenance records will also provide information on the particular requirements and the condition of the saw at the last scheduled maintenance. These records can include the service records and running logs.

So in summary we can see that the more of this information that is available the better. We can determine what is required to be done, what materials may be required, how to isolate and carry out the inspection and testing of the motor circuit and the machine. We can determine a suitable period of time required to carry out the work and discuss the requirements with the client to determine a suitable time period for the work to be undertaken.

Task

Maintenance is to be carried out on an installation which contains a number of machines. List the information that should be available to prepare a programme of work for the maintenance of the machines and the circuits which supply them.

_____

_____

_____

_____

_____

_____

_____

Part 4 Replacement and refitting

When we are undertaking maintenance there is often the need to replace components and it is important that when this is carried out the replacements used are suitable. In some instances it may not be possible to obtain a manufacturer's genuine part and care must be taken when selecting any pattern replacements or alternatives.

Many of the considerations are going to be similar for any electrical maintenance activity so we will consider the common basic requirements.

The first consideration is whether an identical like-for-like replacement is available for the system or component. The like-for-like replacement from the same manufacturer will generally ensure complete compatibility.

There are a number of key issues which must be considered:

Voltage: the replacement component should be of the same voltage rating as the original. There are occasions where the manufacturer increases or decreases the voltage rating and this would need to be checked. We may also need to confirm that the original voltage rating was suitable and not a cause in the need to replace the component.

Current: the current rating or carrying capacity of the equipment must be compatible with the original. Again we need to confirm that the original current rating was correct and not a factor in the need for replacement.

Power rating: the power rating of components such as resistors can be particularly important and replacements should always be in accordance with the manufacturer's specification.

Physical size: when replacing components it is important to make sure that the replacement is of the same physical size as the original.

Duty: the duty of equipment and components needs to be considered as the period of time for which a component or item of equipment is in use and the rate at which the use occurs has an effect on the specification for the replacement.

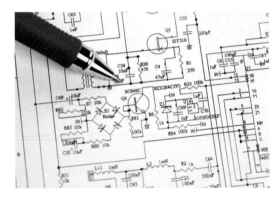

Figure 4.14 *The correct components must be used*

These factors apply to the majority of replacement items used during the maintenance process.

There are considerations to be made depending on the type of system or equipment being maintained. We will consider the various systems and how some of these factors affect the replacement and refitting requirements.

Systems

Three-phase four-wire distribution systems: these are mostly used in the public tertiary distribution system supplying LV installations. Some clients will have their own private distribution system where they purchase energy at high voltage. This is most common on particularly large commercial and industrial sites with a high demand. The particular current and voltage rating of this equipment will be of particular relevance for the cables and equipment. The size of conductors for current carrying capacity and the rating and fault capacity of protective devices must be carefully considered. The type of protective device should also be considered as this will determine their operating characteristics.

Low voltage single and multiphase circuits: the design of the circuits ensures that the installation is safe for use. The conductors, protective devices and accessories replaced during maintenance must comply with BS 7671 and maintain the safety of the users of the installation. The csa and current carrying capacity of the conductors must meet the requirements for the load and voltage drop. Protective devices must be of the correct type and rating with the appropriate short circuit current capacity. Similar requirements will apply for overload devices, isolators, switches and other control equipment and accessories.

ELV: where extra low voltage circuits and equipment is involved the same criteria will apply in terms of the materials and equipment for the installation. The source of the supply needs to be considered and where this requires replacement we need to confirm that the source is as intended. We must take care that where the source is from a separated source (SELV or PELV) that an appropriate source such as a SELV transformer is used.

Lighting systems: Replacement of lamps and luminaires on lighting systems may not always be identical to the original, particularly where energy saving measures are to be introduced. Where the replacement of lamps or control gear is on a like-for-like basis then there are some important considerations to be made. Lamps should be of the same power rating and type, and the colour rendition should be the same. For fluorescent tubes and light-emitting diodes (LEDs) there are a number of colour renditions including warm white, bright white and clean white. Coloured versions are also available and special locations such as plant propagation may require particular light rendition such as daylight lamps.

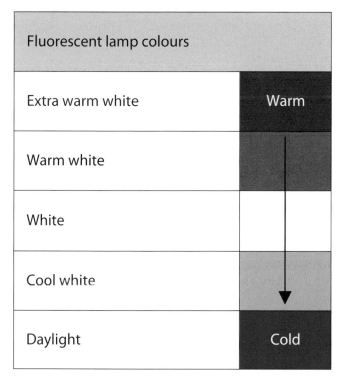

Fluorescent lamp colours	
Extra warm white	Warm
Warm white	
White	
Cool white	
Daylight	Cold

Figure 4.15 *Typical white fluorescent lamp colours*

Heating and ventilating systems: the replacement of components in heating and ventilation systems requires careful consideration in all the general areas such as voltage, current carrying capacity and so on. In addition there are particular control equipment components which need additional care. The setting range for controls and thermostats must be the same as the original components. Thermal overloads and links are essential for the safety of the system and the users of the premises and so they must be of the same type and limits as the original equipment. Motors and fans need to be of the same rating and duty but may be of different manufacture to the original, so a replacement pump may be of a different make but must be the same voltage and power rating with the same duty rating as the original.

Air conditioning and refrigeration systems: basically the same considerations apply to these systems as we considered in the heating and ventilation systems.

Drive systems: drive systems include direct coupling, flexible, gear and belt drives each of which will have particular considerations. Direct couplings may include a shock absorption coupler which contains a compressible material, often rubber, to reduce the shock on the motor and equipment during starting and stopping. These may need to be replaced during maintenance due to wear. It is important that the replacement is suitable for the load which will be applied during the operation of the equipment.

Figure 4.16 *Shaft coupling with shock absorber*

Flexible drives are often used where precise alignment may be difficult to achieve and are subject to the same considerations as the direct drives.

Gear drives may be used to change the ratio between the motor and the equipment such as the raising and lowering of hoists. Where these need to be replaced during the maintenance process it is important to ensure that the load, duty and ratio of the replacement are the same as the originals.

Belt drives come in a number of forms and materials and replacements must be chosen with care to ensure the operation and efficiency of the equipment is not compromised. The choice of belt, such as toothed, flat and vee belts must be the same as the original. The material of the belt must be suitable for the load and duty and should be the same as the original unless the same profile and type can be provided with a better performance specification.

Figure 4.17 *Belt drive*

Security systems: the maintenance of security systems may involve the replacement of sensors, switches and detectors which must be to the same specification as the original. Many of these systems operate at ELV with the supply from batteries which are maintained in a charged state by the normal supply. Any replacement of the supply transformer, rectifier or batteries must also be to the same specification as the original equipment.

Earthing systems: the installation earthing systems must be in accordance with the requirements of BS 7671 and any replacement components on these systems must comply with the current requirements. Over time these requirements have changed and often the existing arrangements may not comply with the current standards. This does not necessarily mean that the installation is unsafe but any replacement work carried out must be to the latest edition of BS 7671. This means that the earthing conductors may need to be installed or the csa of the existing conductors may need to be increased to meet the current requirements. In installations where IT and Data Systems are installed there is often an additional clean earth system.

Protective bonding systems: the same criteria apply to the protective bonding systems installed in the premises. There are also requirements for the protective bonding for data equipment installations to minimize electromagnetic disturbances. The requirements for this are included in Part 4, Section 444 of BS 7671 and these should be complied with.

Data communication: we have considered the particular requirements for the earthing and bonding of data installations above. The standard requirements we identified earlier apply to the data equipment such as voltage, current and duty. The type and location of the cable installed for these systems must be appropriate and suitable for the system and their installation method. The separation of the data cabling from other services is often an issue in older installations and the replacement of the cables may also involve a change of route or containment system.

Figure 4.18 *Solar PV panels*

Solar photovoltaic (PV) supplies: the supply from PV systems is produced by sunlight on the solar panel producing a dc output. This is fed to a control unit where it is converted to an ac output and supplied to the installation. This involves the use of dc cables and isolators on the PV side and ac cables and control equipment on the ac side. Any replacement components must be suitably selected for that part of the system in which they are to be installed. It is important to remember that during daylight hours the output from the solar panels cannot be isolated and so up to the first dc isolator the system will be live. The only method of isolating the supply from the panels is to cover them with a suitable cover. Simply throwing a blanket over them will not normally prevent the generation of dc voltage. Extreme care must be taken when working on the panels and dc cabling.

The requirements for the rating of the solar panel and the component parts of the installation is very much the same as we have already considered for the electrical systems. Care must be taken that any replacement panels are compatible with the existing panels in terms of physical size and output.

Task

1 State five common factors that must be considered when selecting equipment and materials for maintenance.

_____

_____

_____

2 A lighting circuit supplying 230V fluorescent discharge lights is to be maintained. State the information that you would need to confirm before the replacement lamps can be ordered.

_____

_____

_____

We have considered the requirements for types of systems and in doing so we have discussed the requirements for some of the associated equipment including those for:

● Electrical components and accessories
● Motors and starters
● Switchgear and distribution panels
● Control systems
● Luminaires and lamps.

When selecting suitable components for replacements during the maintenance process we need to follow a fairly standard process irrespective of the type of equipment involved.

When selecting the components the procedure will generally involve the following steps:

● Identify the component to be replaced
● Record the details such as voltage, current, resistance, operational limits, ratings, physical dimensions and so on
● Refer to the manufacturer's information or in the case of an installation the design details
● Confirm the item to be replaced meets the specification from the information provided
● Source the supplier of the spare/replacement parts to confirm these are still available
● Source alternatives where original replacements are not available.

It is the second and third items which are specific to the type of equipment being replaced and we considered these requirements earlier in this chapter.

In addition to the requirements for replacement materials and components we have also to identify the resources we need to carry out the work. The most reliable sources for this information are the drawings and maintenance schedules.

We considered the requirements for tools and access equipment in Part 1 of Chapter 2 of this study book and so we will not consider these requirements again here. We can use the layout drawings to determine the access routes and equipment such as steps, ladders, scaffolds and the like. By reference to the maintenance schedules we can identify the tools necessary to undertake the regular maintenance and any specialist tools that may be required.

The general tools and equipment will include:

● Hand tools
● Power tools and whether these are 230V and/or 110V ac and/or battery operated
● Devices used for rotating, straightening and positioning
● Jacks and rams
● Access equipment including ladders, scaffolds and mobile elevated work platforms (MEWPs)
● Portable and fixed lifting equipment
● Trolleys and hand operated trucks
● Fork-lift trucks and cranes.

Figure 4.19 *Select appropriate hand tools*

Materials

There may be some specific material requirements depending on the nature of the maintenance being undertaken. For example:

Cables and insulation: determine whether the conductors are copper or aluminium and if the insulation is thermoplastic or thermosetting. Thermoplastic insulated cables have a lower operating temperature than the thermosetting insulated ones due to the different characteristics of the material. In some locations there is often a requirement for low-smoke and fume (LSF), cross-linked polyethylene (XLPE), halogen free low smoke (OHLS) or fire resistant insulation such as Firetuff or FP200. Any replacement cables should meet the same specification as the original. Should a repair be carried out then the materials used, such as sleeving, must also have the same properties as the original.

Additional protection may be required for cables and insulation in locations where they may be subjected to high temperatures and fibreglass sleeving may be required within luminaires and heating equipment to protect the insulation.

When working on motors and transformers we may need to have a suitable electrical varnish, resin or shellac as this is used to provide insulation. It is commonly used in motor and transformer windings and the control ballasts for discharge lighting. The varnish provides an insulation between the conductors of the windings and takes very little space as the coating is thin. Work on these conductors may damage or require removal of the insulation which will then need to be reinstated.

Figure 4.20 *Shellac insulation on armature windings*

Ceramic and paxolin materials are also used in the construction of electrical equipment as insulators and these may become damaged or need to be replaced during the maintenance process.

Structural and construction metals such as steel, brass and aluminium must be considered as connections and these must also be compatible. For example aluminium conductors cannot be connected directly to brass as this results in an electrolytic reaction between the materials and corrosion takes place.

When working with some systems, such as fibre optic cables and jointing materials, in addition to the specialist tools we may require solvents for cleansing and preparing surfaces. When carrying out soldering on connections within equipment the type of solder flux to be used should be established. The incorrect flux may result in the corrosion of the connection over time and lead to failure.

When considering the drawings, maintenance schedules and manufacturers' instructions these factors must be taken into account to ensure that there is sufficient and suitable material available for the maintenance work to be carried out.

Working conditions

The working and environmental conditions will have an effect on the equipment and components in an electrical installation. It is important to ensure that the equipment and components are suitable for their intended use in the environment in which they operate. The basic requirement for this in terms of the installation and equipment is the use of the IP code (see Table 4.2) to determine the suitability for the location and the environmental hazard. Where the hazards relate to solid bodies, liquids or impact then the IP code is the first reference. The use and implementation of the IP code is covered in the Termination and Connection of Conductors study book in this series.

Table 4.2 *Basic IP code*

IP Code			
1st Digit	**Level of protection**	**2nd Digit**	**Level of protection**
0	Not protected	0	Not protected
1	Protected against solid foreign objects of 50 mm diameter and greater	1	Protected against vertically falling water drops
2	Protected against solid foreign objects of 12.5 mm diameter and greater	2	Protected against vertically falling water drops when enclosure is tilted up to 15°
3	Protected against solid foreign objects of 2.5 mm diameter and greater	3	Protected against water sprayed at an angle up to 60° on either side of the vertical
4	Protected against solid foreign objects of 1.0 mm diameter and greater	4	Protected against water splashed against the component from any direction
5	Protected from the amount of dust that would interfere with normal operation	5	Protected against water projected in jets from any direction
6	Dust tight	6	Protected against water projected in powerful jets from any direction and heavy seas
No code		7	Protected against temporary immersion in water
No code		8	Protected against continuous immersion in water, or as specified by the user
Additional letters			
Additional letter		**Level of protection**	
A		Back of the hand	
B		Finger (12 mm)	
C		Tool (2.5 mm)	
D		Wire (1.0 mm)	

Figure 4.21 *Saw mills and woodworking shops present a potentially explosive atmosphere*

There will be some locations where additional protection is required such as potentially explosive atmospheres such as petrol station forecourts, flour mills and paint spray booths. These will have addition requirements and the persons carrying out work in these areas must be suitably CompEx certified to carry out work in the specific location.

In addition to the requirements for the equipment and installation in the normal operating conditions we must also consider the conditions which exist when the maintenance work is being carried out. Let's consider a few examples.

For example when installing thermoplastic cables if the temperature is around 0 °C then the material becomes brittle and can be damaged during the installation.

The termination of fibre optic cables should not be carried out in a dusty environment, once termination is complete this need not present a problem but during the termination process terminations can be seriously affected.

The insulation material (magnesium oxide) in MIMS cables is hygroscopic (absorbs moisture). Once terminated and sealed the insulation is incredibly durable and robust but termination in damp atmospheres can result in a breakdown in the insulation. This would require re-termination and the moisture would need to be driven from the cable. This can be done by heating the cable with the ends open but is not practical for sheathed MIMS cable.

We can see that the environmental conditions that the installation equipment and components will need to operate in, and the conditions that exist whilst we are working on them, will have an effect on their suitability and lifespan. Indeed in adverse conditions the work can be completely compromised and need to be redone. It is important that we consider all these requirements and select materials, and plan the work, accordingly.

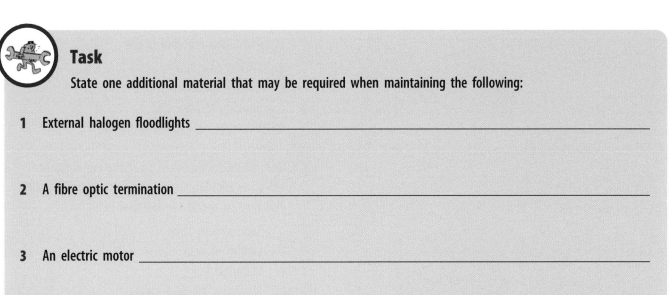

Task

State one additional material that may be required when maintaining the following:

1 External halogen floodlights _____

2 A fibre optic termination _____

3 An electric motor _____

Congratulations you have now finished Chapter 4 of this study book and the end of this unit. Complete the self assessment questions before you carry on to the end test.

SELF ASSESSMENT

1   A client will have some specific requirements when agreeing a programme of maintenance work. Which of the following is **not** one of those specific requirements?

a.   The dates on which work will begin and finish

b.   The supplier of the work equipment and materials

c.   When supplies, circuits or equipment can be isolated

d.   What equipment must be back in service at the end of each day

2   Any changes to the specification must be confirmed by the client:

a.   Once the work is completed

b.   On a deviation report

c.   Using a variation order

d.   By telephone

3   An approved code of practice is:

a.   Statutory

b.   Legislation

c.   Non-statutory

d.   Legally binding

4   When preparing a schedule of work the first item of information required is the:

a.   Wiring diagrams

b.   Client's specification

c.   Supply characteristics

d.   Electrical installation certificate

5   One consideration when selecting replacement cables in addition to the current carrying capacity and type of conductor material is:

a.   Type of insulation

b.   Length of run

c.   Manufacturer

d.   Minimum bending radius

End test

1. **Equipment provided by an employer must be maintained. An employee inspection of the equipment should take place:**

   ☐ a. Prior to use

   ☐ b. Only on new equipment

   ☐ c. When a fault has occurred

   ☐ d. Only every 3 months

2. **Method statements for maintenance activities are not likely to include:**

   ☐ a. The date on which work is to be started

   ☐ b. How the work is to be carried out

   ☐ c. Precisely what is to be done

   ☐ d. Any special requirements

3. **(1) Health and safety at work is the responsibility of every employer and employee**
   (2) Your responsibility for health and safety can be passed to your supervisor

   ☐ a. Both statements 1 and 2 are correct

   ☐ b. Statement 1 is correct and statement 2 is incorrect

   ☐ c. Both statements are incorrect

   ☐ d. Statement 1 is incorrect and statement 2 is correct

4. **Where the work is to be carried out has implications for the health and safety of the operatives and those around them. The activity should be carried out under:**

   ☐ a. A code of practice

   ☐ b. A permit to work

   ☐ c. A method statement

   ☐ d. Increased supervision

5. **The diagram which details the position and function of the components within a circuit is known as a:**

   ☐ a. Wiring diagram

   ☐ b. Layout diagram

   ☐ c. Circuit diagram

   ☐ d. Block diagram

6. **The type of drawing which indicates the particular location of outlets and accessories and Is drawn to scale is known as a:**

   ☐ a. Circuit diagram

   ☐ b. Block diagram

   ☐ c. Wiring diagram

   ☐ d. Layout diagram

7. A common, easily interpreted, method of planning and monitoring progress on-site is by the use of a:

☐ a. Pie chart

☐ b. Bar chart

☐ c. Critical path network

☐ d. Progress register

8. The maintenance requirement for the installation equipment is usually contained in:

☐ a. A data chart

☐ b. A bar chart

☐ c. The servicing records

☐ d. The manufacturer's manual

9. The isolating valve shown in Figure 1 is a:

☐ a. Gate valve

☐ b. T-ball valve

☐ c. Ball valve

☐ d. Lock device

Figure 1

10. When maintenance is required on a construction site or in a factory the preferred voltage above earth potential for handheld portable equipment is:

☐ a. 55 V ac

☐ b. 110 V ac

☐ c. 120 V ac

☐ d. 240 V ac

11. The most suitable method of storing fluorescent tubes prior to their installation would be in a secure container and:

☐ a. Vertical in one corner of the container

☐ b. Laid horizontally on the floor

☐ c. Suspended from the roof

☐ d. Laid horizontally on a shelf

12. The tool shown in Figure 2 is a:

☐ a. Ring spanner

☐ b. Ratchet spanner

☐ c. Open-end Spanner

☐ d. Torque wrench

Figure 2

13. Which of the following access equipment is only suitable for work of a short duration?

☐ a. Low level work platform

☐ b. Extension ladders

☐ c. Mobile elevated work platform

☐ d. Tower scaffold

14. Test instruments used in maintenance work must be:

☐ a. The employee's property

☐ b. Multi-function

☐ c. Drop tested

☐ d. Calibrated

15. When working in an area where several other trades are also working it is important to ensure that the various activities are:

☐ a. Performed in random order

☐ b. Carried out simultaneously

☐ c. Carried out in sequence

☐ d. Coordinated

16. Any pre-existing damage that is identified must be reported to and confirmed with the client:

☐ a. Verbally

☐ b. In writing

☐ c. By telephone

☐ d. On completion of the work

17. When working in an area with fine finish wooden floors, which of the following would prevent damage to them?

☐ a. Covering them with corrugated plastic

☐ b. Covering them in plastic sheets

☐ c. Working without shoes

☐ d. Working on dust sheets

18. The electrical symbol shown in Figure 3 is a:

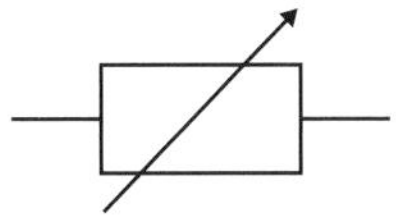

Figure 3

☐ a. Variable resistor

☐ b. Photo cell

☐ c. HBC fuse

☐ d. Diode

19. Of the following documents, which is not statutory?

☐ a. Personal Protective Equipment at Work Regulations (PPE)

☐ b. Provision and Use of Work Equipment Regulations (PUWER)

☐ c. Workplace (Health and Safety and Welfare) Regulations

☐ d. BS 7671 Requirements for Electrical Installations

20. Emergency lighting is provided to allow:

☐ a. Work to be continued as normal

☐ b. Work to be continued with care

☐ c. Evacuation of the premises

☐ d. Access by the emergency services

21. Self-contained luminaires are required to be maintained:

☐ a. Three yearly

☐ b. Annually

☐ c. Six monthly

☐ d. Monthly

22. Which of the following categories is the correct one for fire protection for properties?

☐ a. M

☐ b. L1

☐ c. P1

☐ d. FPP1

23. The maximum period generally given for the overhaul of switchgear is:

☐ a. Annually

☐ b. Five yearly

☐ c. Ten yearly

☐ d. 15 yearly

24. **The client has requested a change to the programme of work to enable a valuable order to be fulfilled. This will require the work to be completed 2 days earlier than originally programmed. To achieve this the options are to:**
    i   **Increase the number of maintenance staff on-site**
    ii   **Work outside normal working hours**

    Which of the following statements would be the advice given to the client?

    ☐ a. Action i) only will incur additional cost

    ☐ b. Action ii) only will incur additional cost

    ☐ c. Neither action i) or ii) will incur additional cost

    ☐ d. Both actions i) and ii) will incur additional cost

25. **Maintenance that is required when pre-set parameters are measured and the levels set are reached is called:**

    ☐ a. Non-routine maintenance

    ☐ b. Monitored maintenance

    ☐ c. Breakdown maintenance

    ☐ d. Planned preventative

26. **Changes to the specification for the work to be carried out are usually recorded on a:**

    ☐ a. Variation order

    ☐ b. Confirmation order

    ☐ c. Contractor's order

    ☐ d. Change to specification order

27. **Where maintenance is required which has not been scheduled or demanded by failure it is known as:**

    ☐ a. Preventative maintenance

    ☐ b. Breakdown maintenance

    ☐ c. Condition monitored maintenance

    ☐ d. Non-routine maintenance

28. **By it's definition an electrical system consists of a source of supply, connecting conductors and:**

    ☐ a. An installation

    ☐ b. A current-using device

    ☐ c. A transformer

    ☐ d. A battery

29. **The British Standard BS 1363 relates to:**

    ☐ a. A 13 A socket outlet

    ☐ b. A three-plate ceiling rose

    ☐ c. The fuse fitted in a 13 A plug

    ☐ d. An Edison screw lampholder

30. **A component has been identified as needing to be replaced and a pattern spare part is to be used. The considerations for the suitability of the replacement will include the:**

    ☐ a. Manufacturer

    ☐ b. Power rating

    ☐ c. Cost

    ☐ d. Delivery

Safe isolation

5

The material contained in Chapters 5 to 9 covers the knowledge required for C&G Unit No. 2351–302 (ELTK 09) and the equivalent EAL Unit.

Carrying out and ensuring safe isolation is an essential part of an electrician's work. For this reason the requirements for safe isolation appear in a number of the units of the national occupational standard. This chapter considers the requirements for safe isolation in relation to electrical maintenance work. In doing so it includes additional information to that contained in the other Study books.

LEARNING OBJECTIVES

On completion of this chapter you should be able to:

● Specify the correct procedure for completing safe isolation

● State the implications of not carrying out safe isolations to:

- Self

- Other personnel

- Customers/clients

    – Public

    – Building systems (loss of supply).

● State the implications of carrying out safe isolations to:

    – Other personnel

    – Customers/clients

    – Public

    – Building systems (loss of supply).

Part 1 The requirement for safe isolation

This chapter considers the requirement for safe isolation of electrical circuits and installations to enable electrical maintenance work to be carried out safely.

Whilst working through this Chapter you will need to refer to Health and Safety Guidance Note GS 38, Electrical test equipment for use by electricians.

During the course of our electrical maintenance work there are many occasions where we will be required to work on a circuit which is in service. This is normal to allow us to maintain the circuit or the equipment it supplies. It is important to ensure that the part of the installation, circuit or equipment we are going to work on is safely isolated from the supply. Safe isolation does not simply mean making sure that the supply is switched off; it also includes making sure that it is not inadvertently re-energized.

Figure 5.1 *Safe isolation is required*

There are a number of reasons why safe isolation must be carried out and this action will have implications for building systems, other people and ourselves. Similarly failure to carry out safe isolation will also have its implications and perhaps it would be as well to consider these first.

Failure to safely isolate

The most obvious implication of the failure to safely isolate when working on electrical installations and equipment is the risk of electric shock to ourselves and others.

The effect of electric current on the bodies of humans and animals is well recorded. The values quoted here are generic and so should be taken as general guidance. A current across the chest of a person in the region of 50 mA (0.05 Amperes) and above is enough to produce ventricular fibrillation of the heart which may result in death. As the average human body resistance is considered to be in the region of 1 KΩ (1000 Ω) with a voltage of 230 V then the current would be in the region of 0.23 A (230 mA). That is over four-and-a-half times the level needed to cause ventricular fibrillation.

When working on the electrical installation or equipment we often have to expose live parts, which if not safely isolated, pose a serious risk of electric shock. Other people and livestock within the vicinity of our work will also be able to access these live parts and may not have sufficient knowledge and understanding to avoid the dangers involved.

Where our work is carried out in public areas this risk is further increased as the installation and equipment may be accessed by anyone; adults, children and animals and the failure to safely isolate presents a very real danger.

Figure 5.2 *Failure to isolate may have serious consequences*

Electric shock also carries the danger of electrical burns which occur at the entry and exit points of the contact and within the body along the path taken by the current. These burns can be severe and whilst the casualty may survive the electric shock the damage, some of which may be irreparable, can be considerable.

To ensure these dangers are removed safe isolation of the installation, circuit or equipment is essential.

The failure to isolate can also affect the building and structure. Failure to isolate introduces a risk of arcing where live parts are exposed. This may occur between live parts at different potentials (line to neutral and between line conductors) and between live conductors and earth.

When an arc is produced electrical energy is converted into heat energy and the level of discharge energy results in molten conductor being present in the arc. This presents a real risk of fire and isolation of the circuit(s) by operation of the protective device will not extinguish a fire started in this way.

In the case of a fault, including one which does not result in an electric shock or fire, the supply should be disconnected automatically. In this case the circuit or installation may be disconnected unexpectedly. The building systems may be switched off resulting in loss of data, failure of heating or ventilation, lighting and power.

In severe cases we may also lose the building and life protection systems such as fire alarm systems, sprinklers, smoke vents, firefighters' lifts and the like. It may also result in the loss of lighting and ventilation to internal areas that have no natural light or ventilation. Lifts, electric doors and escalators may also be affected leaving people trapped or stranded.

These disruptions of supply may also result in considerable expense to the client and damage to the electrical equipment and buildings. The loss of lighting and the other services can cause other dangers to persons within the building resulting in trips, falls and injuries from machinery and equipment.

Safe isolation

Carrying out safe isolation is essential to safeguard against the dangers we have identified above but there are a number of implications which must be considered.

Before we isolate we need to consider the effect this will have on people, the building and the equipment and services. We need to determine the extent of the installation which needs to be isolated to carry out our work safely. There are certain activities which will need the complete installation to be isolated and others where the isolation of one or more circuits may be all that is necessary.

Figure 5.3 *Safe isolation will affect others*

The isolation of the complete installation has serious implications for the users of the installation as the electrical equipment and lighting will not be available for use. This means that the timing and duration of the work must be carefully considered and discussed with the user to minimize disruption.

This will also affect our activities as there will be no supply available for lighting or power tools. So we would need to consider task lighting and power for our work and it may be necessary to arrange a temporary supply for some of the client's equipment.

Where safety services or alarms may be affected we need to consider the consequences of isolation. For example burglar alarms may be linked to the police or a security company and the loss of supply may result in a response visit. If this is a false alarm the client may be charged for the wasted visit. Similarly a fire alarm isolated from the supply may cause an alarm to be triggered which could cause the fire service to respond.

In addition there are a number of other building services which may be affected including time locks, door access systems, bar code readers and tills, security cameras and public address systems.

Isolation of individual circuits may also cause inconvenience to the client and the requirements will need to be discussed to ensure that any disruption is kept to a minimum. In any event we must always obtain permission before we isolate.

In certain instances it may not be possible to carry out some or all of the work during normal working hours. In which case arrangements will need to be made with the client to ensure that access is permitted. This may involve the client in the provision of staff to attend, for security or access purposes, during the period when the work is carried out.

Task

Referring to Health and Safety Guidance GS 38 list the particular requirements for:

1 Test leads

_____
_____
_____
_____
_____
_____
_____

2 Test probes

_____
_____
_____
_____
_____
_____
_____

Try this

1   List three risks which are removed by safe isolation.

a _____

b _____

c _____

2   Electrical maintenance work is to be carried out in a small baker's shop. List three client needs that should be considered when circuits are to be isolated to allow work to progress.

a _____

b _____

c _____

Part 2 Safe isolation procedures

The key to safety is to follow the correct procedure throughout the isolation process. We will first consider the process for safe isolation of a single circuit so we can work on it safely.

Let's consider a situation where we are to carry out maintenance on a lighting circuit in a shop. Before we can begin to work on the circuit it must be safely isolated from the supply. This circuit is protected by a BS EN 60898 type C circuit breaker (cb) in the shop's distribution board located at the origin of the installation.

First we must get permission to isolate the circuit. This must be obtained from the person responsible for the electrical installation (the Duty Holder), **not** just any employee. As we are going to be isolating the supply the duty holder must ensure that the safety of persons and the operation of the business is not going to be compromised. To do this the area to be affected and the duration of the isolation should be explained to the duty holder to help in making the decision.

Once we have been given permission to isolate the circuit we must correctly identify the particular circuit within the distribution board. Where there are a number of lighting circuits it is important that we isolate the right one. Providing the distribution board has been correctly labelled and the appropriate circuit charts are available this should be relatively straightforward.

There are proprietary test instruments which allow the identification of a circuit before it is isolated. These rely on being connected to the circuit when they then transmit a signal through the circuit conductors. A second unit is used to sense the signal at the distribution board to identify the fuse or circuit breaker. The sensitivity of the unit can be adjusted to give a clear and reliable indication of the circuit to be isolated.

This works well with circuits which include a socket outlet as the sender unit can be readily plugged into the circuit. However where the circuit does not include a suitable socket a connection would need to be made to live parts.

This introduces a higher risk to the operative as, due to the purpose of the device, the circuit will not be isolated. In these circumstances the operation would require two persons, one to operate the sensing unit and the other to make the connection to the circuit using test probes complying with GS 38. Extreme care is required when accessing the live terminals to make this connection and it should only be carried out by skilled persons using suitable equipment.

Having correctly identified the circuit the cb is switched off, isolating the circuit from the supply.

An appropriate locking off device is then fitted to the cb to prevent unintentional reenergizing of the circuit. There are a number of proprietary devices available for this task and they all perform the same function, which is, preventing the operation of the cb. In most cases a separate padlock is inserted through the locking off device to prevent unauthorized removal. This secures the cb and a warning label should also be fitted to advise that the circuit should not be energized and that someone is working on the circuit. Typical lock off devices and labels are shown in Figures 5.5 to 5.7.

Courtesy of Martindale Electric

Figure 5.5 *Typical warning label*

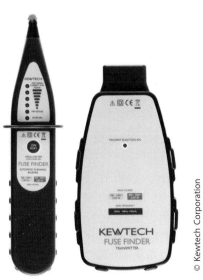

© Kewtech Corporation

Figure 5.4 *Typical circuit identification instruments*

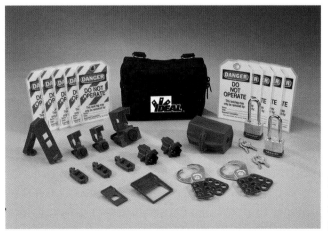

Ideal Industries

Figure 5.6 *Typical lock off kit*

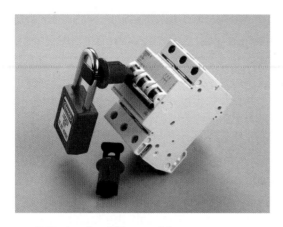

Figure 5.7 *Lock off in position*

Having safely isolated and locked off the circuit we must now confirm that the intended circuit is actually isolated from the supply. To do this we will need an approved voltage indicator together with a proving unit, as shown in Figure 5.8. The term Approved Voltage Indicator refers to a voltage indicator which meets all the requirements of Health and Safety Guidance GS 38.

Figure 5.8 *Typical proving unit and approved voltage indicator*

In this instance we are going to remove the cover from a ceiling rose on the circuit at the point where we are going to start work and confirm that it is actually isolated. The first step, having removed the cover, is to confirm that the approved voltage indicator is functioning using the proving unit. The output produced by the proving unit should cause all the light-emitting diode (LED) indicators to light showing that they are all functioning correctly.

As this is a single phase lighting circuit we will need to confirm isolation by testing between:

● All line conductors and neutral (loop line terminal and switch line terminal)
● All line conductors and earth (loop line terminal and switch line terminal)
● Neutral and earth.

And there should be no voltage present at any of these connections.

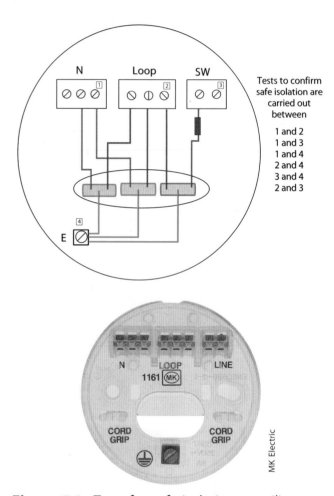

Figure 5.9 *Tests for safe isolation at ceiling rose*

Finally we must confirm that the approved voltage indicator is still working. To do this we will use the proving unit again and all the LED indicators should light when the approved voltage indicator is tested.

This process will have confirmed that the circuit we are going to work on is isolated from the supply.

However it is always advisable when removing accessories to carry out work to further check at each accessory that isolation is achieved. It may be that a luminaire or accessory which appears to be on the circuit is actually supplied from elsewhere.

For example it is not uncommon to find a downstairs socket wired from an upstairs circuit and once the downstairs circuit is isolated it would be logical to assume that all the ground floor sockets are isolated. Similarly where two circuits supply the same area, such as the ground floor of a dwelling, then it is not always obvious which sockets are supplied on which circuit.

Summary procedure for circuit isolation:

1   Seek permission to isolate
2   Identify the circuit to be isolated
3   Isolate by switching off cb or isolator
4   Fit locking device and warning label
5   Secure area around accessory to be removed (barriers)
6   Remove accessory
7   Select an Approved Voltage Indicator (AVI)

8   Confirm AVI complies with GS 38
9   Confirm the operation of the AVI using a proving unit
10 Test between all live conductors
11 Is circuit dead? If not go back to 2
12 Test between all live conductors and earth
13 Is circuit dead? If not go back to 2
14 Confirm AVI is functioning using proving unit.

Isolation of a complete installation follows a similar procedure but in this case we are isolating the supply to the whole installation or all the circuits supplied from a particular distribution board.

When isolating a number of circuits it is important to discuss with the duty holder the areas that will be isolated and determine any special requirements related to any of the circuits which will be isolated.

If we are to confirm safe isolation of a three phase distribution board then there will be a number of additional tests to be carried out as identified in Figure 5.10.

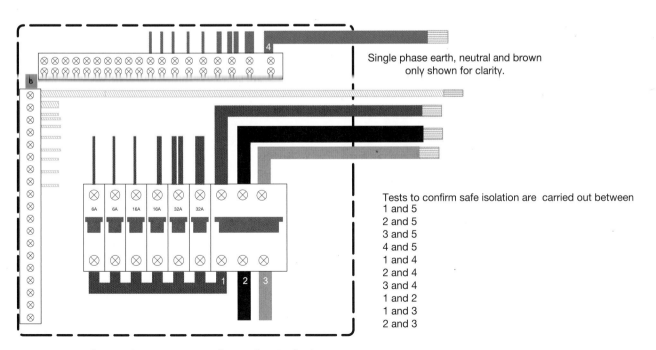

Single phase earth, neutral and brown only shown for clarity.

Tests to confirm safe isolation are carried out between
1 and 5
2 and 5
3 and 5
4 and 5
1 and 4
2 and 4
3 and 4
1 and 2
1 and 3
2 and 3

Figure 5.10 *Three phase points of test for isolation*

Summary procedure for distribution board isolation:

1   Seek permission to isolate
2   Identify the distribution board to be isolated
3   Isolate by switching off main isolator
4   Fit locking device and warning label
5   Remove distribution board cover to access live terminals
6   Select an Approved Voltage Indicator (AVI)
7   Confirm AVI complies with GS 38
8   Confirm the operation of the AVI using a proving unit
9   Test between all live conductors
10 Is circuit dead? If not go back to 2
11 Test between all live conductors and earth
12 Is circuit dead? If not go back to 2
13 Confirm AVI is functioning using proving unit.

Note

A safe isolation flow chart is included at the end of this chapter for your reference. You can copy this and put it with your test equipment as an aide-memoire.

Remember

When we consider proving the operation of the AVI we can use a known live supply or a proving unit. When isolating a distribution board it is possible to use the incoming supply to the isolator to prove the AVI is functioning before and after we test for isolation. This is not possible when isolating a single circuit and so a proving unit is essential for circuit isolation.

Task

Look around your workplace, training centre or local supermarket and identify any equipment which would require special consideration if the installation were to be isolated from the supply.

_____

_____

_____

_____

_____

_____

_____

_____

_____

You have now completed Chapter 5. Correctly complete the following self assessment questions before you carry on to the next chapter.

SELF ASSESSMENT

1   One implication of carrying out safe isolation is:

   a.   Increased shock risk
   b.   No inconvenience to the user
   c.   The installation will function normally
   d.   The use of the installation will be restricted

2   One implication of not carrying out safe isolation is:

   a.   Increased shock risk
   b.   Inconvenience to the user
   c.   The installation will not function normally
   d.   The use of the installation will be restricted

3   The equipment used to confirm safe isolation of supply is:

   a.   An Approved Voltage Indicator
   b.   An ordinary bi-pin lamp
   c.   A continuity tester
   d.   A multimeter

4   Isolation is only confirmed once tests have been made between:

   a.   All line conductors
   b.   All live conductors and earth
   c.   All live conductors and all live conductors and earth
   d.   All line conductors and neutral and all line conductors and earth

5   Once isolation has been confirmed the test equipment must be confirmed to be:

   a.   Compliant with GS 38
   b.   Working, using a proving unit
   c.   Suitable for the expected voltages
   d.   In calibration

Safe isolation flow chart

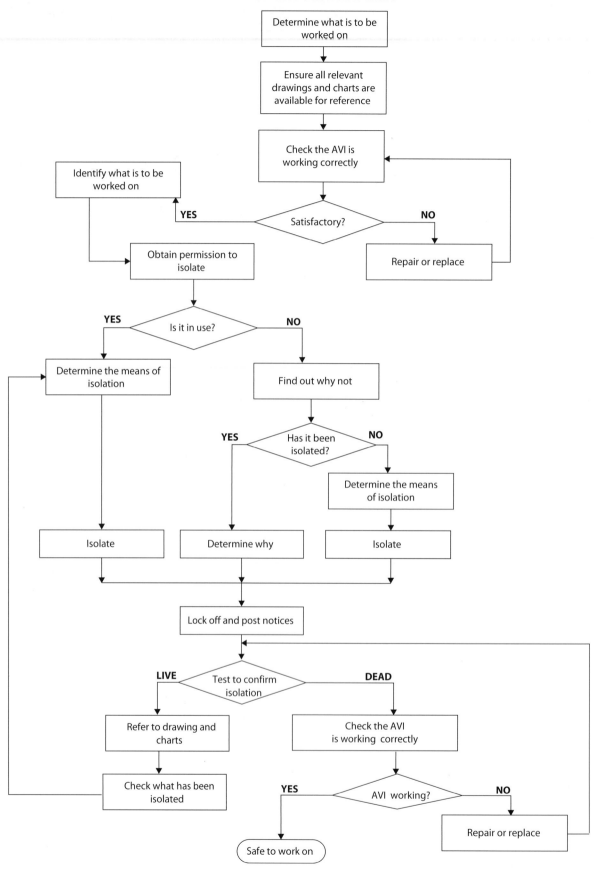

Consumer supply systems

6

RECAP

Before you start work on this chapter, complete the exercise below to ensure that you remember what you learned earlier.

Safe isolation involves ensuring the _____ is switched _____ and it is _____ so that it is not inadvertently _____.

Failure to safely_____ when working on electrical installations and _____ means a serious _____ of electric shock to _____ and others.

Electric shock also carries the danger of electrical _____ which occur at the _____ and _____ points of the contact and _____ the body along the _____ taken by the current.

Before isolating, the effect of this action on _____, the building, the equipment and _____ must be considered.

Permission to isolate must be _____ from the person _____ for the electrical installation not just any employee.

To confirm that a single phase circuit is safely isolated tests are carried out between:

● All _____ conductors and _____

● All _____ conductors and _____

● _____ and _____

and there should be _____ voltage _____ at any of these connections.

LEARNING OBJECTIVES

On completion of this chapter you should be able to:

- Explain the characteristics and applications, of the following systems:

  - Earthing arrangements:

    - TN-S

    - TNC-S

    - TN-C

    - TT

    - IT.

  - Supply systems:

    - Single phase

    - Three phase

    - Three phase and neutral.

- Specify the arrangements for electrical installations and systems with regard to provision for:

  - Isolation and switching

  - Overcurrent protection

  - Earth fault protection.

This chapter considers the characteristics and application of consumer supply systems and the arrangements for isolation, switching and protection. There will be a need for you to refer to BS 7671 when you are working through this chapter.

Part 1 Earthing arrangements

Before we consider the earthing arrangements we need to understand why and how we earth the electrical supply system.

The general mass of earth is generally accepted as a conductor of electricity and is at a potential of zero volts. As our electrical installations operate at voltages above zero, there is a potential difference between the mass of earth and our electrical system. In order to control any current flow which may occur to the mass of earth, we need to connect our system to earth. To achieve this electricity substations have the neutral point connected to earth.

It is also a requirement that all electrical installations in general use have a connection to this Earth. The consumer's Earth connection may be provided by the Distribution Network Operator (DNO) by using the metal sheath of the supply cable or a separate conductor within the supply cable. The system using this type of provision is referred to as a TN-S system.

Alternatively the Earth provision may be provided by use of the neutral conductor within the supplier's network, which is then referred to as a protective earth neutral (PEN) conductor, and this system is referred to as a TN-C-S or protective multiple earthing (PME) system.

Where a consumer's Earth connection is not provided by the supplier then a separate installation earth electrode must be installed to provide the connection to Earth. This is referred to as a TT system and it is normal practice to fit a residual current device (RCD) to protect the installation.

We shall be considering these public supply systems in more depth later in this chapter.

Earth faults

Earth faults are generally caused by live parts coming into contact with exposed metalwork

which is then made live. In order to prevent this dangerous situation from arising we connect the exposed metalwork to earth. The reason for this is to provide a safe return path for earth fault currents.

The earth fault path

The purpose of the earth fault path is to:

- Allow fault currents to return safely to the supply transformer and for them to
- Disconnect the supply to the faulty circuit before any danger from fire, shock or burns can occur.

Let's look at the earthing arrangement for each system in more detail. For clarity we will use a simple single phase supply circuit for each system diagram which will not include things like the energy meter, distribution board, consumer unit and so on.

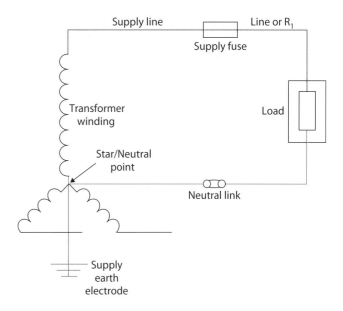

Figure 6.1 *The basic supply circuit with no earthing arrangement shown*

A system is a single source of supply and an installation and Figure 6.1 shows the basic circuit taken from a three phase transformer. We shall consider each earthing system in terms of a single winding of the supply transformer, supply cables between the transformer and the installation and the installation itself.

As we saw earlier each type of system earth has a particular classification which we shall use for their identification.

TN-S system

This tells us that:

1st letter **T** – the supply is connected directly to earth at one or more points

2nd letter **N** – the exposed metalwork of the installation is connected directly to the earthing point of the supply. (The neutral of the supply system is normally earthed.)

3rd letter **S** – a separate conductor is used throughout the system to provide the connection of the exposed conductive parts to the earth of the supply.

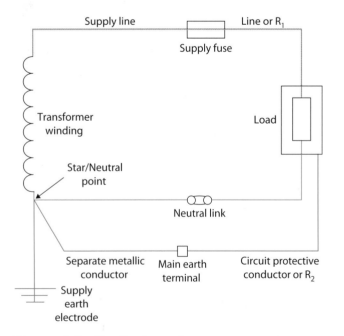

Figure 6.2 *The TN-S earthing arrangement*

This earth connection is usually either through the sheath of the supply cable or a separate conductor within the supply cable. Separate protective conductors are used within the installation to connect the exposed conductive parts to the main earthing terminal. In the event of a fault to earth in this system the current flow will be around the earth fault loop as shown in Figure 6.3.

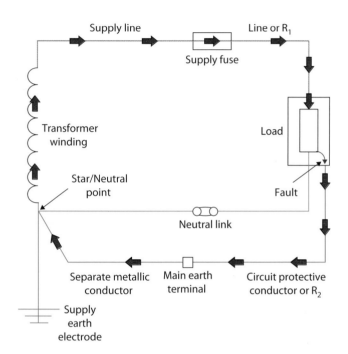

Figure 6.3 *TN-S Earth fault path*

As a conductor is used throughout the whole system to provide a return path for the earth fault current, the return path should have a low value of impedance. The protective conductors are generally a smaller cross-sectional area (csa) than the live conductors.

TN-C-S system

This type of system is similar to the TN-S system except for one important feature as we shall see.

1st letter **T** – the supply is connected directly to earth at one or more points

2nd letter **N** – the exposed metalwork of the installation is connected directly to the earthing point of the supply. (The neutral of the supply system is normally earthed.)

3rd letter **C** – within the DNO's supply system the function of neutral conductor and earth conductor are combined in a single common conductor known as a protective earth neutral (PEN) conductor.

4th letter **S** – a separate conductor must be used throughout the installation to provide the connection of the exposed conductive parts to the main earthing terminal. The use of a combined neutral and earth conductor within the installation is **not** permitted.

Again as electrical conductors are used throughout the system a low earth fault return path impedance should be obtained. In the event of an earth fault on this system the current flow will be around the earth fault loop as shown in Figure 6.4.

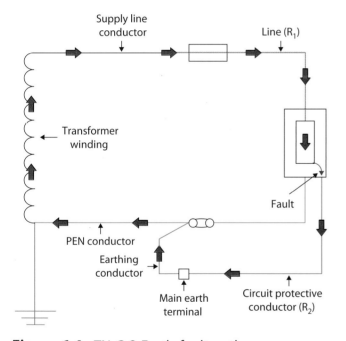

Figure 6.4 *TN-C-S Earth fault path*

TT system

This tells us that:

1st letter **T** – the supply is connected directly to earth at one or more points

2nd letter **T** – the installation's exposed metalwork is connected to earth by a separate installation earth electrode.

The connection between the supply and installation electrodes, and therefore the return path for earth fault current, is the general mass of earth. When a fault to earth occurs on this system, the earth fault current will flow around the circuit shown in Figure 6.5.

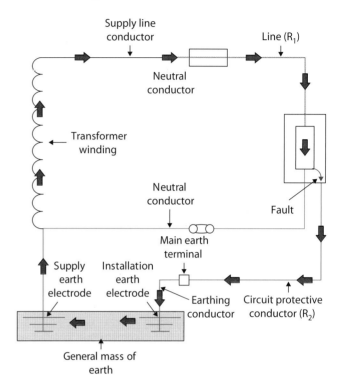

Figure 6.5 *TT Earth fault path*

In this system, the earth fault current returns through the general mass of earth which will generally have a high value of impedance.

The path through which the earth fault current flows is known as the earth fault loop path and the earth fault loop impedance plays a major part in the protection against electric shock.

> **Remember**
>
> A current flow of 0.05 A, that is 50 milliamps, is enough to cause a fatal electric shock to people. So protection against even very small currents is vital to prevent danger from electric shock.

These three systems are the types available on the public distribution system. There are two other systems of which we should be aware. These are the TN-C and IT systems which are not available on the public distribution network and are not suitable for general use.

TN-C System

The TN-C earthing arrangement is rarely used and it is one where a combined PEN conductor fulfils both the earthing and neutral functions in both the supply and the installation.

IT System

In the IT system there is either no connection of the supply to earth or the system has only a high impedance connection and an insulation monitoring device monitors the impedance.

The TN-C and IT systems require special precautions and the supervision of a skilled person when in use.

As stated earlier the impedance of the earth fault path plays an important part as it will regulate the amount of current that flows in the event of a fault to earth. This earth fault loop impedance value should be measured to confirm compliance when maintenance is undertaken.

The system earth fault loop impedance (Z_s) comprises the external earth fault loop impedance Z_e, which is the DNO's part of the system, and the resistance of the line (R_1) and the circuit protective conductor (R_2) of the electrical installation and so $Z_s = Z_e + (R_1 + R_2)$.

The measurement of the earth fault loop impedance is carried out using a special test instrument and is measured between line and earth. We need a low value of impedance to ensure a good return path to encourage large currents to flow when a fault to earth occurs.

Remember

- The main function of the earthing in each system is to provide a safe return path for any earth fault currents which may occur.
- The impedance of the total current route taken under earth fault conditions is known as the 'earth fault loop impedance'.

Try this

1   The initial letters for the TN-C-S system refer to:

   a   T

   _____

   b   N

   _____

   c   C

   _____

d S

_____

2 Describe how the earth return path between the electrical installation and the supplier's transformer is formed for the following systems

a TN-S

_____

_____

_____

b TN-C-S

_____

_____

_____

c TT

_____

_____

_____

Part 2 Supply systems

Distribution systems

The parts of the distribution system which are closest to the consumer's intake are those which operate at:

- 11 kV called the secondary distribution
- 400/230 V called the tertiary distribution

with the tertiary distribution being the most common for supplies up to 100 A.

The Electricity Suppliers have a legal responsibility to keep the supply within the limits identified in the Electricity Safety, Quality and Continuity Regulations. Following voltage standardization in Europe these are:

- For voltage a nominal supply of 400/230 V, + 10%, – 6%
- That frequency must not be more or less than 1% of 50 Hz over a 24-hour period.

The public distribution system can be roughly split into three main consumer groups:

- Industrial
- Commercial and domestic
- Rural.

Industrial consumers may take their supply at 33 kV and in some cases 132 kV. Where an industrial estate consists of smaller units, the estate will have a substation supplied with say 33 kV or 11 kV. The transformer will then stepdown the voltage to the tertiary distribution at 400/230 V for supply to the units.

Large commercial premises may have their own substation transformer fed at 11 kV, which will step down the voltage to 400/230 V for internal distribution, whilst smaller commercial and domestic consumers are usually supplied at 400/230 V.

The 11 kV input to the transformer will be connected in delta whereas the 400/230 V output will be a star arrangement (Figure 6.7). To supply the delta connected windings a three-phase three-wire system is used, with no neutral conductor. The star connected output uses a three-phase four-wire connection with the centre point of the star being the neutral which is connected to earth.

Try this

Using Figure 6.6 insert the most appropriate voltage in each of the boxes.

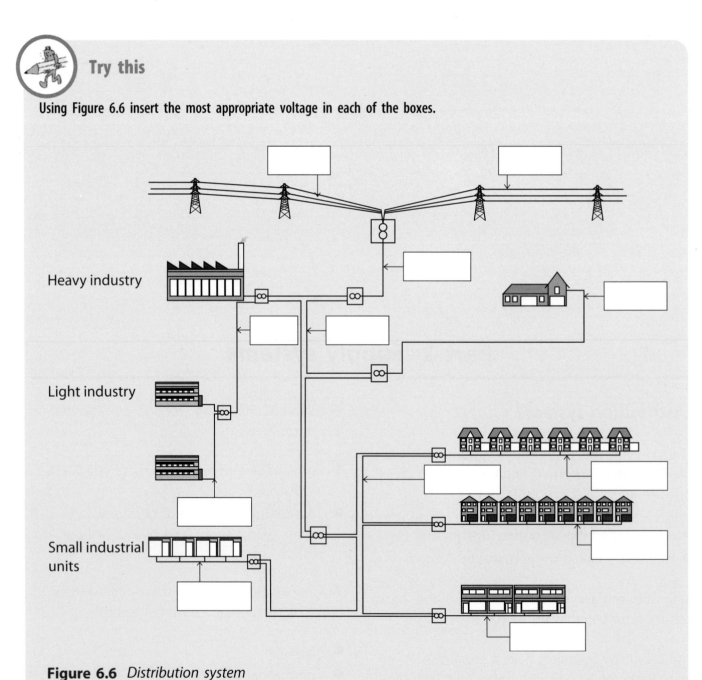

Figure 6.6 *Distribution system*

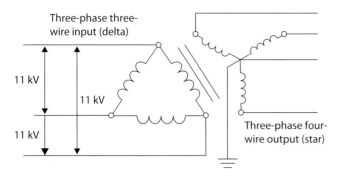

Figure 6.7 *Typical delta/star transformer arrangement*

In delta the voltage across each of the lines (line voltage) is the same as the transformer winding, whereas a transformer winding connected in star will give us a voltage between lines of 400 V (U) and a voltage between any line and the neutral star point of 230 V (U_o), as shown in Figure 6.8

The number and colour identification of line conductors in the UK is Brown/L1, Black/L2 and Grey/L3 with the neutral Blue/N.

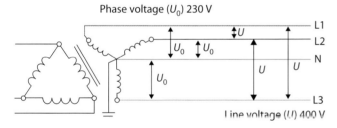

Figure 6.8 *Voltages available from a star connected transformer winding*

The relationship between the voltage line to neutral and the voltage line to line for a star connected winding is:

$$\text{Voltage Line to Neutral} = \frac{\text{Voltage between Lines}}{\sqrt{3}}$$

$$\text{Voltage Line to Line} = \text{Voltage Line to Neutral} \times \sqrt{3}$$

Where $\sqrt{3}$ = approximately 1.732

> ✏️ **Note**
>
> The line to line voltage is equal to the sum of the two line to neutral voltages each multiplied by Sine 120° (voltages displaced by 120°). The Sine of 120° is 0.866 which is equal to half $\sqrt{3}$. As there are two L-N voltages displaced by 120° we have the line to neutral voltage x $\sqrt{3}$.

Example:

The nominal line to neutral voltage is 230 V for most single phase installations on the public network in the UK. The voltage between lines for this supply is:

$$
\begin{aligned}
230 \times \sqrt{3} &= 230 \times 1.73 \\
&= 400 \text{ V}
\end{aligned}
$$

Similarly a 400 V voltage between lines gives a voltage line to neutral of:

$$\frac{400}{\sqrt{3}} = \frac{400}{1.73} = 230 \text{ V}$$

Try this

For star connected windings calculate the line to line voltages if the line to neutral voltages are:

1   400 V _____

2   220 V _____

3   120 V _____

Load currents in three-phase circuits

It is important to recognize the relationships of the currents in star and delta connected windings. In star connected the current through the line conductors is equal to that flowing through the phase windings, as shown in Figure 6.9 and so we can see that $I_L = I_P$.

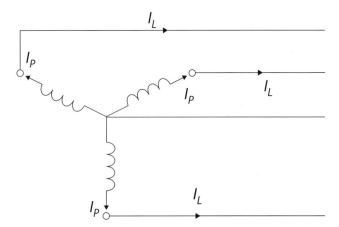

$I_P = $ Phase winding current
$I_L = $ Line conductor current

Figure 6.9 *Current in a star connected load*

However in the delta connected winding this appears to be more complex. The line current, when reaching the transformer winding, is split into two directions so that two-phase windings are each taking some current.

As each of the phases is 120° out of phase with the others, and the current is alternating, each line conductor acts as a flow and return. The waveforms in Figure 6.10 show that due to this 120° displacement there is current flowing in the positive and negative half cycles at all times. As these are connected together through the windings all the currents are the same, the current will effectively cancel each other and so there

will be no need for a neutral conductor in a delta connected load.

Remember

- In a single-phase system the neutral will carry the load current
- In a star connected three-phase system the neutral conductor will carry the out of balance load current.
- In a balanced three-phase system (delta connected) there is no imbalance and so a neutral conductor is not required.

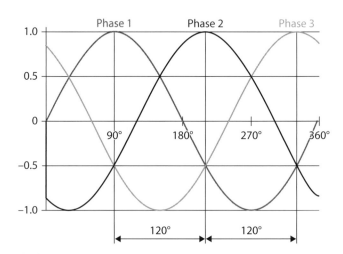

Figure 6.10 *Balanced three-phase current waveform*

For example: At 90º Phase 1 positive = Phase 2 + Phase 3 negative so the arithmetic sum is zero

Considering what we have learnt about current in a balanced delta system the arrows shown on Figure 6.11 only give an indication as to the current distribution for each line. All of these currents would **not** be flowing in the directions shown at the same time.

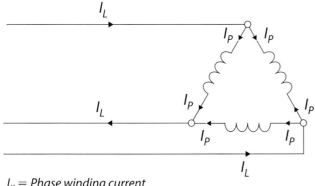

I_P = Phase winding current
I_L = Line conductor current

Figure 6.11 *Current in a delta connected load*

The currents through the phase windings are:

$$I_P = \frac{I_L}{\sqrt{3}}$$
$$or \; I_L = I_P \times \sqrt{3}$$

Example:

If the line current is 100 A the phase winding current is:

$$I_P = \frac{100}{\sqrt{3}} = \frac{100}{1.73}$$
$$= 57.8 \text{ A}$$

Three phase balanced loads

All transmission and primary distribution is carried out using a three-phase system. It is important that each of the phases carries about the same amount of current.

Three-phase motors have equal windings and each phase is the same. Therefore the conductors carry the same current, and these automatically create a balanced load situation.

For domestic areas the output of the star connected transformer is 400/230 V. All premises are generally supplied with a single phase and neutral at 230 V. To try to balance the loads on each of the phases, houses may be connected as shown in Figure 6.12. If it was possible to load all of the phases exactly the same the current in the neutral would be zero.

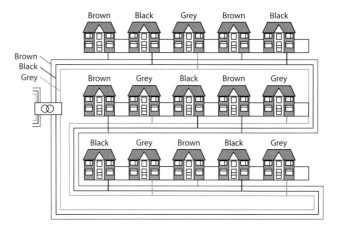

Figure 6.12 *Houses are connected so that their loads are spread across the three phases*

The distribution of electrical energy is tailored to the needs and location of the consumer in relation to the supply source. The use of transformers to determine the distribution system voltage allows the supplier to size cables and control gear to give maximum network efficiency and convenience to the customer. When the supply enters the consumers' premises connection is made to the customer's own distribution system and final circuits. These systems and circuits will vary depending on the type of consumer, the energy requirement and the type of equipment being used, so we shall consider typical final circuit arrangements for the most common groups of consumers.

Try this

Describe the function of the neutral conductor in a:

1   Single-phase circuit

_____

_____

_____

2   Three-phase star circuit

_____

_____

_____

3   Balanced three-phase delta

_____

_____

_____

Domestic

Domestic consumers generally receive their supply at 230 V, single-phase, from the supply company, although in areas where electric space heating is employed a two- or three-phase supply may be provided. This is because, depending on the size of the heating load, it may be necessary to distribute the current demand across the phases.

The final distribution circuits for the domestic installation, lighting and power circuits, operate at 230 V, single phase throughout. Some items of equipment, for specific locations or in order to provide specialist control or effects, are supplied at a lower voltage. These items of equipment will generally operate at extra-low voltage (ELV), below 50 V ac, and are supplied via their own local transformer.

Figure 6.13 _SELV lighting in a kitchen_

Each final circuit must be controlled by a protective device suitably rated to supply the load of the circuit which it protects. The requirements for the protection of the final circuits are covered later in this chapter. Each final circuit must have its own protective device and the distribution board in domestic installations is generally a consumer unit containing miniature circuit breakers or fuses. Other protective devices, such as RCDs may also be incorporated.

A typical domestic consumer unit is shown in Figure 6.14, and the rating of the devices will vary between say a 6 A device for the protection of the lighting circuits through to 45 A devices for the larger power using equipment such as large electric showers and cookers.

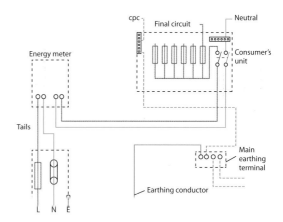

Figure 6.15 *Typical TN-S domestic schematic diagram*

Commercial

The arrangement of final circuits in small commercial premises may be very similar to that of the domestic installation. However, larger commercial buildings will have a considerably bigger floor area and a greater power requirement. The same consideration needs to be given to the economic use of conductor sizes and effective utilization of equipment within the consumer's installation as is given to the supply network.

As a result it is common to install distribution circuits within the installation. In a multi-storey building, for example, a distribution circuit may be installed to each floor. These would generally comprise a means of isolation and protection at the intake position, a large cross-sectional area cable to a convenient point on the appropriate floor and a distribution board containing the final circuit protection devices for the circuits on that level.

MK Electric

Figure 6.14 *Typical (dual RCD) domestic consumer unit*

It is common to try to locate these distribution boards as close to the centre of the area as possible in order to minimize the length of the final circuit cables. It is common for the consumer's distribution circuit to be at 400 V, three-phase and neutral with the lighting and power circuits for the floor area being supplied at 230 V single-phase. Again special areas, or the requirement for sophisticated controls, may require ELV systems to be installed.

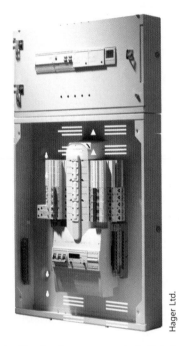

Figure 6.16 *Typical commercial distribution board*

A similar approach may be taken to a single storey building which covers a large floor area with the distribution boards being placed at convenient locations around the building. These distribution board locations are often referred to as power or load centres.

Industrial

Small industrial units will often receive their supply at 400/230 V from the supply company and

their final circuit arrangements may be very similar to those of the commercial installation above.

Many industrial applications involve the use of equipment with a high power requirement, which is often achieved most economically with three-phase equipment. It is quite common to find that the general lighting and power in such installations are at 230 V single-phase and that 400 V three-phase final circuits are installed to supply particular items of equipment such as motors.

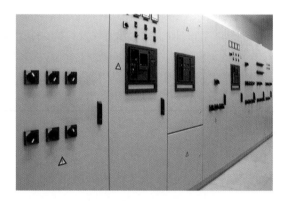

Figure 6.17 *Large three-phase industrial system*

We can categorize the basic final circuits into some specific areas and consider the requirements for each.

Heating

The circuits in this category will include those supplying space heating, water heating and cooking applications within the installation. As a general rule these final circuits will be supplying equipment which will draw considerable current, and require large cross-sectional area cables to deliver it. By supplying such circuits at voltages greater than 230 V single phase, 50 Hz, the current required to produce the same power output can be considerably reduced. Installations which receive their supply at 400 V and above may benefit from supplying their large heating loads at 400 V three-phase.

Power

General power outlets are provided at 230 V and these are common to most installations. The type of outlet may vary depending on the type of equipment and the intended use of the circuit, but the BS 1363, 13 A socket outlet appears in almost every installation. Larger equipment which requires more power will be supplied by a dedicated final circuit. Some industrial applications require the installation of 400 V, three-phase or three-phase and neutral socket outlets. These are often used to supply portable or transportable equipment which would otherwise require a much larger single-phase circuit.

Lighting

Lighting is generally supplied at 230 V single phase, and the type of lamp used determines the light output, with specialist lamps being employed for applications such as street lighting, car parks and sports halls. Some applications require the lighting to be supplied at ELV and some locations may require this to be separated extra-low voltage (SELV). Both of these operate

at no more than 50 V ac. The difference between them being that the SELV is supplied through a safety isolating transformer to BS EN 61558-2-6.

Figure 6.18 *Street lighting*

Control circuits

Many electrical installations now include equipment which requires sophisticated control circuitry. The domestic dimmer switch, operating at 230 V, 50 Hz, is perhaps one of the most basic controls. Industrial and commercial installations use equipment which often requires an elaborate control system. The commercial installation may require controls on air conditioning and heating systems and there may also be a need for environmental control within an installation. Many companies now use energy management systems which monitor and control all aspects of energy consumption. These systems, whilst seemingly expensive, can in the longer term offer considerable cost savings for the company. The industrial consumer will often require complex control over the production process resulting in considerable savings in both manpower and reduced wastage.

Figure 6.19 *Industrial control panels*

These control systems often involve the use of microprocessors and logic controls and whilst the equipment being controlled may have a high power consumption the actual control processes require minute power levels. The current requirements are very low and as a result the use of ELV equipment in electronics is most suitable.

With voltages as low as 12 V ac or dc and the small current requirement means that small cables can be used for the interconnection of the control devices. Control circuits are generally supplied via panel-mounted transformers for the ac devices, or transformer rectifiers for the dc equipment. The control circuits are often supplied through special devices to prevent fluctuations in the supply or interference from the supply system interfering with the control system.

Alarm systems

Alarm systems also operate on very small power requirements, as they basically rely on a change of state in an electronic circuit to cause the operation of an alarm. The majority of systems use electronic equipment throughout and, like our control systems, operate at voltages as low as 12 V.

Many systems are available, some operating on ac and some on dc. Generally the supply to the system is derived from a 230 V ac supply and the equipment transforms and rectifies this to provide the ELV for the alarm system. Many alarms operate through internal supplies which incorporate a device allowing the system to be supplied from a separate, usually self-contained, dc source. This source, normally a battery, is maintained in a fully charged state by the main supply. However in the event of a main supply failure the battery is able to take over and run the system.

Figure 6.20 *Domestic alarm control panel*

Try this

1 **List four typical voltages that may be found in an electrical installation within an industrial unit which is in the manufacturing industry stating their use in each case.**

_____

_____

_____

_____

_____

2 State the typical supply voltages for each of the following installations:

a domestic dwelling _____

b commercial premises _____

c large industrial premises _____

Part 3 Protection

The main requirements

There are a number of requirements that must be considered when dealing with control and protection of an installation, the main ones being:

- Isolation and switching
- Protection against overcurrent
- Earth fault protection.

All these requirements are related to safety and need to be considered for protection, not just from electric shock, but also from fire, burns or injury from mechanical movement of equipment which is electrically activated.

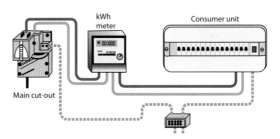

Figure 6.21 *The main intake position of a domestic installation*

Within the installation other local methods are used and the positioning and type of these can be very important.

Isolation and switching

The term isolation, within this context, means the cutting off of the installation, or circuit, from all sources of electrical supply to prevent danger. In a domestic installation the main means of isolation is usually the main switch controlling the consumer unit.

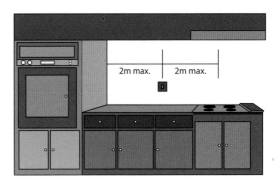

Figure 6.22 *Cooker isolator*

The control for a cooker is a double pole switch which should be positioned no further than 2 m from the cooker or, as in Figure 6.22, from either part of the cooker, but not above it. We use similar isolation arrangements for immersion heaters and domestic boilers and the like.

In an industrial installation the requirements can become far more complex. Whilst within a domestic installation most isolators are double-pole types, in industrial situations there will not only be double pole isolators there will also be triple-pole and triple-pole and neutral (TP & N).

Some examples of the use of isolators in industrial installations are shown in Figures 6.25 and 6.26. Figure 6.27 is an example of emergency switching.

Figure 6.23 *Immersion heaters must have a double pole control switch adjacent to the heater*

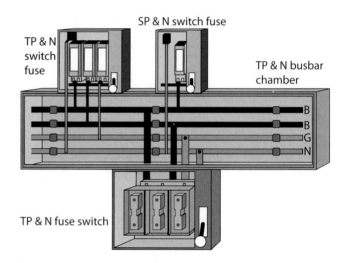

Figure 6.25 *Industrial distribution*

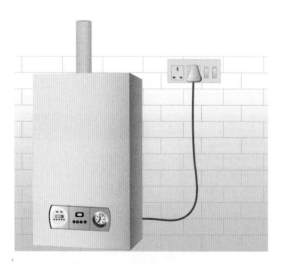

Figure 6.24 *Domestic boilers often have a plug and socket as the means of isolation.*

Figure 6.26 *Motor with an adjacent isolator*

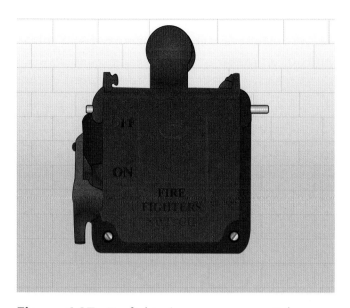

Figure 6.27 *Firefighter's emergency switch*

> **Remember**
>
> An isolator must cut off an electrical installation, or a part of it, from every source of electrical energy.

Overcurrent protection

Overload

An overload is a situation that occurs in a circuit which is still electrically sound. It is generally caused by trying to take more power from a circuit than it is designed for which results in a larger than normal current flowing in the circuit. If the load is reduced then the circuit can continue to function without any need for repair.

A typical example of this is a radial circuit which has been extended and the load has increased to exceed that originally intended. Each item connected in the circuit and the circuit supplying the equipment is healthy but more current is being drawn through the cable than was originally intended.

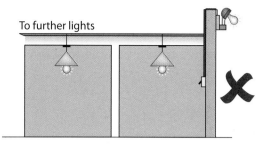

Figure 6.28 *A radial circuit with ten existing lights has a 500 W tungsten halogen lamp added which increases the load to exceed the circuit capacity*

Every cable has some resistance, and the result of drawing more current through the cable is the production of more heat in the cable. This rise in temperature will, over a period of time, result in the insulation becoming less effective and eventually breaking down. In the case of severe overload the insulation becomes so hot it begins to melt and may even catch fire. This is obviously a serious fire risk and we must take steps to prevent this happening.

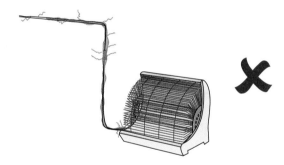

Figure 6.29 *Severe overload creates a real fire risk*

Adding load to a circuit over time, such as plugging in additional equipment to a socket outlet circuit, can result in a gradual increase in load current which may continue for some time before the protective device disconnects from the supply. So an overload can be a gradual increase in current.

Short circuit

A short, unlike an overload, is the connecting together of live conductors which results in a very rapid increase in current, in nanoseconds. This produces a very high current in the circuit, as the only impedance in the circuit is that of the electrical conductors and this is very low. The protective device has to be able to safely disconnect the potential current that would flow in the case of a short circuit, and this may be in thousands of amperes.

Try this

Explain how:

1   **An overload may occur in a circuit.**

_____

_____

_____

2   **A fire may be started by an overloaded cable.**

_____

_____

_____

Part 4   Earth fault protection

Earth fault protection is designed to protect against electric shock, fire and burns as a result of a fault to earth. This will result in all the exposed and extraneous conductive parts within the installation becoming live.

The installation we are considering is part of a TN-S system and the installation is connected to the supply transformer with line and neutral conductors and the earth is provided by the supply cable sheath.

An appliance, in this case an electric kettle, is plugged into a socket outlet. The element will be connected between line and neutral to make it work normally, and the metal case will be connected to the earth pin of the plug. Under normal operating conditions no current would flow in the circuit protective conductor.

However, if a fault develops, such as corrosion of the element's outer cover, resulting in current leaking from the live element through the water to the metal case of the kettle, then the case of the kettle becomes live. This creates a real risk of electric shock and the kettle must be disconnected before it can create a danger.

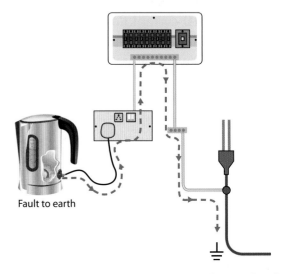

Fault to earth

Figure 6.30 *Fault to earth on an electric kettle*

As the metal case of the kettle is connected to earth, the fault current will flow through the earth return path and back to the earthed point of the supply transformer. There has to be a complete circuit for current to flow. By creating a low impedance return path using good electrical conductors a complete circuit for the return of the earth fault current to the source transformer is provided.

Remember

We are using impedance for the earth fault loop circuit because it is an ac current which will flow in the circuit.

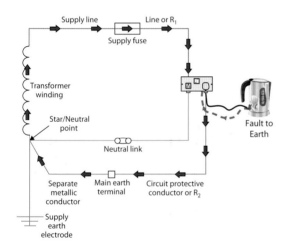

Figure 6.31 *Fault path in the TN-S system*

It is important that the impedance of the earth fault path is kept at as low as possible so that in the event of an earth fault the maximum current can flow. We need this high current to flow in order to operate the protective device and disconnect the supply.

There are two significant requirements here:

● The fault is disconnected from the supply
● The disconnection is fast enough to prevent a fatal electric shock occurring.

If the earth fault path has a high enough impedance the current may continue to flow and the equipment remain live without the protective device operating.

Remember

The I_n (current rating) of a protective device is the current it can carry for an indefinite time without deterioration.

For a protective device to operate current in excess of the I_n rating must flow. For example: a 32 A BS 88-3 type fuse will take around 140 A to cause the fuse to disconnect in 5 seconds and 240 A to disconnect in 0.4 seconds. For final circuits rated up to and including 32 A BS 7671 requires a disconnection time of 0.4 seconds to provide fault protection.

We can see from this that if only a small earth fault current occurs, due to the high earth fault loop impedance, then the protective device will not operate in time.

A current of around 50 mA (0.05 A) is sufficient to cause ventricular fibrillation of the heart for the average person, which can result in death. If a fault current of 50 mA flows to earth on a circuit protected by a protective device rated at say 10 A the current in the line conductor is only 10 A + 50 mA, a total of 10.05 A and this will not cause the protection device to operate.

It is important to remember that the human body can be affected by currents as low as 5 mA. If the fault path passes through the human body there is a risk of electric shock even at these very low current levels.

Note

Appendix 3 of BS 7671 contains time current curves for protective devices and each set of curves contains an inset table which indicates the current required for the protective device to operate within given times.

Earth fault impedance test

To ensure that this earth fault path has a low enough earth fault impedance an earth fault loop impedance test is carried out. This test passes a current through the earth return path from the premises to the supply transformer, through the windings and back to the consumer's premises through the line conductor.

There are two tests involving earth fault loop impedance:

- Z_e: the earth fault loop impedance of the earth return path external to the installation (the supplier's part of the system)
- Z_s: the earth fault impedance of the whole system (the external path and the line and circuit protective conductor (cpc) conductors of the circuit).

Z_e is measured at the origin of the installation with the installation isolated from the supply and all parallel paths disconnected. This is normally done by disconnecting the earthing conductor from the main earthing terminal, hence the need for the installation to be isolated from the supply.

Z_s is carried out at the furthest point of all radial circuits (the point of maximum impedance for the earth fault loop path). The test is also carried out at all accessible socket outlets on ring final circuits.

Note

There is more information on testing, including testing earth fault loop impedance in the study book 'Inspection, Testing and Commissioning' in this series.

Remember

If the earthing arrangement for an installation becomes disconnected exposed metalwork of the installation (exposed conductive parts) and metalwork of other services and the building structure (extraneous conductive parts) may become live and reach the supply voltage U_0 of 230 V.

Task

Determine, using the information contained in Appendix 3 of BS 7671, the current required to cause each of the following protective devices to operate within the time given.

1   16 A, BS 88-3 fuse system C, in 0.4 seconds _____

2   45 A, BS 3036 semi-enclosed fuse, in 0.2 seconds _____

3   100 A, BS 88-2 fuse system E, in 5 seconds _____

4   40 A, BS EN 60898 type B circuit breaker in 0.4 seconds. _____

 **Try this**

1   Explain why a low earth fault loop impedance is necessary to provide fault protection.

_____

_____

_____

_____

_____

2   Using the information in this chapter list all the component parts of the earth fault loop impedance path in a TN-S system starting at the point of fault.

_____

_____

_____

_____

_____

Crossword

Across

1	Cables supported on pylons are here (8)
4	Adding equipment to a circuit may cause this (8)
8	This voltage is extra low but all alone (4)
10	This may become live during an earth fault (9)
11	Fuses and a raincoat can offer this (10)
13	This system has a separate earth throughout (3)
15	This system has an installation electrode (2)
16	Used to prevent the risk in the event of a fault to earth (8)
17	How the energy is spread to consumers (12)
19	Fault paths must have a low one (9)
20	A form of 18 down (3)

Down

2	Out in the open or can be touched (7)
3	The main isolator for a single phase installation is this (6-4)
5	They are responsible for the distribution system (3)
6	This type of connection has no neutral (5)
7	To stand on a tightrope and be 6 down needs to be (8)
9	Turning on or off (9)
12	Positively not required for 6 down (8)
14	This twinkles and needs a 12 down connection (4)
18	This system has a PEN conductor (4)

Congratulations you have completed Chapter 6 of this Study book. Complete the self assessment questions before you continue to Chapter 7.

SELF ASSESSMENT

1   The type of supply system which uses the general mass of earth to connect the installation earth electrode to the supply transformer is:

a. TT

b. TN-S

c. TN-C-S

d. IT

2   The current in the phase winding of a large, delta connected, three phase load is 130 A. The current in the line conductor will be:

a. 53 A

b. 75 A

c. 225 A

d. 480 A

3   A fault between line conductor and exposed conductive parts will result in:

a. overload

b. short circuit

c. earth leakage currents

d. no change in circuit conditions

4   To cause the fuse in a lighting circuit to operate when a fault to earth occurs the current which flows in the earth path must be a:

a. high dc current

b. low dc current

c. low ac current

d. high ac current

5   Z_s represents the earth fault loop impedance that is:

a. external to the installation

b. for the complete system

c. within the installation

d. for a single circuit

7

Health and safety requirements for completing maintenance

RECAP

Before you start work on this chapter, complete the exercise below to ensure that you remember what you learned earlier.

The total path taken by earth fault currents is called the earth _____ loop impedance and the _____ of this path must be low to provide _____ protection.

The connection between the main _____ terminal and the DNOs transformer for a TN-C-S system is the_____ conductor and for a TT system it is the _____.

_____ distribution is the system which operates at _____ kV. _____ distribution is the system which operates at _____/230 V.

In a_____ three phase load the current flowing in the _____ conductor will be _____.

The final circuit for domestic installation, lighting and power circuits, operates at _____V, _____ phase.

If the voltage between line conductors of a three phase star connected load is 400 V the voltage across each phase winding will be 400 V _____ by _____.

If the current in each phase winding of a delta connected load is 25 A the current in the line conductor will be 25 A_____ by _____ .

A fault between _____ conductors produces a _____ circuit whilst an increase in load in a healthy circuit is an _____ .

LEARNING OBJECTIVES

On completion of this chapter you should be able to:

● State the procedures for carrying out an assessment of risks and implementing safe systems of work for the completion of maintenance activities

● Identify the appropriate health and safety regulations which apply to work activities, and the persons who are legally responsible for health and safety

● State how to:

– Select and use appropriate PPE

– Select and use appropriate tools and equipment for specific maintenance jobs, including:

– Hand tools

– Power tools (110 V ac or battery operated)

– Portable and fixed lifting equipment

– Access equipment

– Rotating, positioning and straightening devices

– Jacking devices and rams

– Trolleys and hand-operated trucks.

● Select and use appropriate materials for specific maintenance jobs, including:

– Materials for plant, equipment and components for use within maintenance programmes – thermoplastic, thermosetting (rubber compounds), fibreglass sleeving, varnish (shellac), ceramics, metals (conductors, structural), solvents.

● Identify inappropriate work practices and state the implications if such practices are employed.

Part 1 Risk assessment

Whilst on-site you will need to ensure that the equipment and procedures you follow are in accordance with health and safety requirements. The first of these procedures that needs to be considered is a risk assessment.

A risk assessment is carried out to determine whether work activities or the environment and location are likely to cause harm. The objective is to identify and reduce the risks where it is 'reasonably practicable' to do so. Reasonably practicable is a term used in legislation and it means that an employer must balance the cost of the actions taken to reduce the risk against the degree of risk which exists. This will include the trouble, effort and time taken to carry out the task and not just the expense.

Figure 7.2 *Risk assessment form*

The risk assessment should consider everyone who may be affected by the activity, not just those carrying out the work. This will include the client's employees, other contractors, members of the public and visitors to the site.

The Management of Health and Safety at Work Regulations requires employers and the self-employed to carry out a risk assessment for their work and anyone who may be affected by it. The risk assessment should be reviewed periodically and where there are five or more employees the risk assessment and the findings must be recorded. In addition to the general working activities there are a number of specific areas of legislation which also require a risk assessment of specific areas to be carried out. These include:

● The Manual Handling Operations Regulations
● Control of Substances Hazardous to Health Regulations 2002

Figure 7.1 *Risk assessments*

- The Personal Protective Equipment at Work Regulations 1992
- The Control of Asbestos at Work Regulations 2002
- The Display Screen Equipment Regulations.

The risk assessment must identify the actions taken to reduce the risks and it is these actions that you need to confirm are effective and being followed.

Most companies will have a standard risk assessment for their everyday activities which will involve the same risks wherever they are undertaken.

So how is a risk assessment carried out? Well that will vary from one organization to another with each having a slightly different approach. What is important is that a risk assessment is carried out and all the foreseeable risks are considered.

To make a risk assessment it is important to understand the difference between a risk and a hazard. A general description is accepted as:

- **Hazard:** a hazard is anything with the potential to cause harm, so electricity, noise, dust and so on are typical examples of hazards.
- **Risk:** a risk is the likelihood of a hazard causing injury, damage or loss and how severe the outcome may be.

The Health and Safety Executive (HSE) recommends that a risk assessment is carried out in five stages which are:

1 Identify the hazards associated with the work activities
2 Identify who could be harmed by those hazards
3 Identify how you manage the risks at present and what further steps might be

required to reduce the risks further. (These are the control measures.)
4 Record the findings of the assessment and inform those at risk of the controls
5 Review the Risk Assessment on a regular basis, e.g. if the staff, the activity, or the equipment used change.

Figure 7.3 *Assess the hazards and determine the risk*

Each activity should be considered quite critically including how it is carried out. Reference to any existing guidance and information such as accident reports may also be used to help determine the risk.

The risk assessment should be reviewed at least annually.

Remember

We all make risk assessments every day. Before we cross the road we make a risk assessment for the presence, speed and distance of any vehicles and determine whether we are able to cross safely. We make many similar risk assessments when driving a car and risk assessments at work are an extension of this process.

We have mentioned control measures and these are the ways in which we can reduce the risks. To decide on the most appropriate measures we would consider the following points:

- Can the risk be eliminated altogether? So for our electrical maintenance work the risk could be eliminated if it is possible to isolate the equipment, circuit or installation from the supply

- Can the risk be contained by additional procedures or equipment? So maintaining equipment which is installed at high level could be carried out using a number of options. For example we could consider using purpose-built scaffolding or mobile work platforms. Any other associated risks with these systems would also need to be considered such as suitable access, safety harnesses and the need for training

- Can the work be adapted or arranged to suit a particular situation? Where work is to be carried out in areas where high levels of noise exist, such as in a machine shop, one consideration would be to arrange for the work to be carried out whilst there is no work which produces high levels of noise being carried out

- Is there a technological or engineering alternative to carrying out the work? One example of this is the use of a remote control robotic camera to carry out inspections in areas which are particularly hazardous such as high temperature or explosive risk areas.

Once the methods of control have been considered and implemented there may be additional requirements such as additional training. The provision of personal protective equipment (PPE) may also be considered but this should only be as a last resort when there are no other control measures which can be used.

A risk assessment form would be produced which identifies the activity, the hazards, the nature of the risk and the control measures required to deal with them.

DM Associates Ltd Risk Assessment						No.	
Assessed by		Date		Authorized by		Date	
Work activity:							
Task	Hazard	Likely harm	Risk rating	Control measures required	Additional requirements	Further information	

Figure 7.4 *Typical risk assessment form*

When you receive a risk assessment the information may be in a format similar to that shown in Figure 7.4 and you will need to interpret the requirements to ensure they are complied with. Typical risk assessment information when, for example, carrying out live electrical testing would be:

Work activity: live electrical testing when undertaking maintenance work

Task: undertaking live testing

Hazard: contact with live parts, electric shock, short circuit, falls as a result of electric shock

Likely harm: electric shock to operative or other people, electrical burns, and injury from falls – all possibly fatal

Risk rating: high (if uncontrolled), low (if controlled)

Control measures: working in accordance with requirements of DM Associates Ltd 'Live testing guidance', equipment to be in accordance with HSE Guidance GS 38.

Additional requirements: only DM Associates Ltd quality assured and calibrated instruments to be used.

Further information: further guidance can be found in:

- HSE Guidance GS 38,
- IET Guidance Note 3 Inspection and Testing
- Memorandum of Guidance on the Electricity at Work Regulations HS (R) 25.

We can see that the risk assessment has set out the nature of the task, the likely hazards and the possible harm that could result.

The level of risk is given in the Risk Rating and without the control measures the risk is high. If the control measures are implemented the risk is reduced to low.

The control measures are then given and it is here that the need for a level of competence and use of correct procedures and equipment are identified. This may require a certain level of training and demonstrable competence by the operative.

To ensure that operatives are carrying out this activity safely you would need to confirm that they are aware of these requirements. You should also have an understanding of the requirements so that you can confirm the work is carried out in accordance with them.

The guidance material identified in the risk assessment should be available to those on-site.

Try this

The fluorescent lighting in a small workshop is installed on metal trunking at high level through the workshop. The luminaires are to be cleaned and maintained out of normal working hours when the workshop is not in use.

List four activities and the associated hazards which would need to be addressed in a risk assessment before this work can be carried out.

_____

_____

_____

_____

_____

_____

_____

_____

_____

Part 2 Health and safety

Note

You could find it useful to look in a library for copies of the publications mentioned in this part. Read the appropriate sections and be on the lookout for any amendments or updates to them.

Keeping everyone safe whilst at work is the responsibility of both the employer and the employee.

Figure 7.5 *Cooperating but not working safely!*

There are a number of Acts of Parliament and Regulations that govern what employers provide in a workplace and how the employees use what is provided. Both the employers and the employees are legally required to observe safe working practices.

Health and Safety at Work (etc.) Act 1974

The aim of this Act is to improve or maintain the standards of Health, Safety and Welfare of all those at work. It applies to everybody who is at work and it sets out what is required of both employers and employees.

There are a number of regulations and codes of practice which have been introduced under the Health and Safety at Work (etc.) Act, including:

- Personal Protective Equipment at Work Regulations (PPE)
- Control of Asbestos at Work Regulations
- Work at Height Regulations
- Management of Health & Safety at Work Regulations
- Control of Substances Hazardous to Health (COSHH) Regulations
- The Electricity at Work Regulations
- Manual Handling Operations Regulations
- Workplace (Health and Safety and Welfare) Regulations
- Provision and Use of Work Equipment Regulations (PUWER)
- Display Screen Equipment at Work Regulations.

Remember

Watch out for new, and amendments to the existing, legislation regarding Health and Safety.

The employer's responsibility

The Management of Health & Safety at Work Regulations places a duty on employers to provide and maintain a working environment for their employees which is, as far as practicable, safe and without risk to health.

The 'working environment' applies to all areas to which employees have access. For example, corridors, staircases and fire exits are included, as are gangways, stairs and steps.

The 'safe working environment' includes such factors as:

- A clean and tidy workplace
- Suitable and adequate lighting
- Adequate ventilation and fume and dust control
- Maintaining a reasonable working temperature and humidity.

Figure 7.6 *Access routes should be kept clear*

The employer is responsible for working with any other employers or contractors who are sharing their workplace to ensure that everyone's health and safety is safeguarded.

Other facilities that are required by law include those for washing and sanitation which must also be suitable for disabled employees.

First aid

The supply of adequate first aid equipment and facilities is also the employer's responsibility. The contents of a first aid box should be based on the employer's assessment of the particular needs for the company and its activities.

Figure 7.7 *Locate the first aid station*

Any used items should be replaced and any sterile items that have an expiry date must be renewed within their useful life. Sterile water for eye irrigation is required in many locations and this must be stored in a sealed container. Once the seal has been broken the container should not be reused as the water will no longer be sterile and any remaining contents should be disposed of.

Figure 7.8 *First aid box*

It is important to note where the first aid box or first aid station is located and who the appointed first aiders are. The first aid facilities are identified by signs with a white cross on a green background.

Work equipment

Employers are required to provide and maintain suitable, safe tools and equipment for use by their employees. Where necessary, training in the use of such equipment must be provided by the employer. Any information or supervision that may be required is also the employer's responsibility.

Work equipment should be suitable for its intended use and for the conditions in which it is to be used. It should be inspected periodically and maintained in a safe condition to ensure that it remains safe for use. Records should be kept of the inspections carried out on the equipment.

The employer must also ensure that the method of working is safe. Where it is required protective equipment must also be provided.

The storage, handling and transporting of goods is also the responsibility of the employer. Goods should be stacked on suitable shelves and in a manner that will prevent danger. Some materials, especially chemicals, should be stored in the correct containers and labelled clearly. Heavy items may require mechanical handling aids such as a pallet truck for safe transportation.

Figure 7.9 *A pallet truck*

Under the Display Screen Equipment at Work Regulations your employer will need to analyze any workstations and assess and reduce risks. Employers have to provide training in the use of a video display unit (VDU) and workstation so that employees know how best to avoid health problems.

Some common problems, such as aches, pains and tired eyes can be avoided by having a comfortable workstation and taking frequent short breaks.

Figure 7.10 *The workstation*

Reporting accidents

Employers must have insurance, known as an Employer's Liability Insurance, in place in case employees are injured or become ill through work. A current insurance certificate must be readily available either as a printed or electronic copy.

In the event of an accident, however slight, employers must ensure that the details are recorded in an accident book which is kept at the workplace.

Dangerous occurrences, major injuries and deaths must be reported to the relevant enforcing authority. This may be either the local authority Environmental Health Department or the Health and Safety Executive who may wish to investigate further. They may also inspect the company's accident book at the same time.

Major injuries include those that result in:

● Injury from electric shock or burns leading to unconsciousness or requiring resuscitation
● Fractured bones
● Loss of sight or
● Hospitalization for more than 24 hours.

Dangerous occurrences that are reportable include:

● The overturning of a fork-lift truck for example
● An electrical short circuit or overload accompanied by fire or explosion which has the potential to cause death or which results in the stoppage of the plant involved for more than 24 hours.

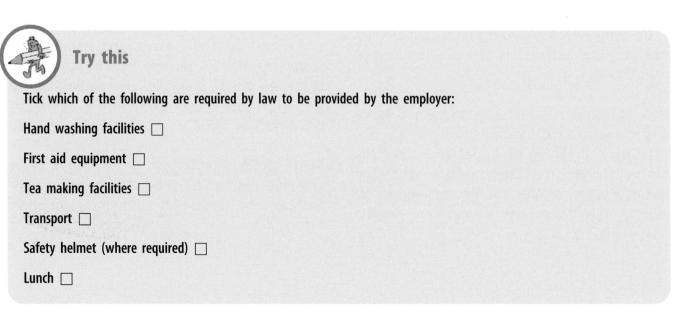

Try this

Tick which of the following are required by law to be provided by the employer:

Hand washing facilities ☐

First aid equipment ☐

Tea making facilities ☐

Transport ☐

Safety helmet (where required) ☐

Lunch ☐

Work-related diseases, such as asbestosis which may be caused by working with asbestos dust, must also be reported. Every accident at work involving an employee, which results in the employee being unable to work for three or more consecutive days, must be notified to the Environmental Health Department or Health and Safety Executive.

Safety policy

It is a legal requirement for an employer who has five or more employees to produce a safety policy and to make it available to all employees. This is normally in the form of a written notice, but it can be stored on a computer. However it is kept it must be available for reference by all staff.

The safety policy is produced following a risk assessment which is carried out by the employer. The safety policy will include details of any hazards and risks present in the working environment together with the safety procedures which need to be taken in order to protect the health and safety of all persons concerned. The safety policy should be subject to regular review in consultation with the safety representatives.

Health and safety policy

This is the statement of general policy and arrangements for:	Name of organisation
Overall and final responsibility for health and safety is that of:	Name of employer
Day-to-day responsibility for ensuring this policy is put into practice is delegated to:	

Statement of general policy	**Responsibility of** ***(Name / Title)***	**Action / Arrangements** ***(Customise to meet your own situation)***
To prevent accidents and cases of work-related ill health and provide adequate control of health and safety risks arising from work activities		
To provide adequate training to ensure employees are competent to do their work		
To engage and consult with employees on day-to-day health and safety conditions and provide advice and supervision on occupational health		
To implement emergency procedures - evacuation in case of fire or other significant incident. You can find help with your fire risk assessment at: (See note 1 below)		
To maintain safe and healthy working conditions, provide and maintain plant, equipment and machinery, and ensure safe storage / use of substances		

Health and safety law poster is displayed:				
First-aid box and accident book are located: Accidents and ill health at work reported under RIDDOR: (Reporting of Injuries, Diseases and Dangerous Occurrences Regulations) (see note 2 below)				
Signed: (Employer)		Date:		
Subject to review, monitoring and revision by:		Every:		months or sooner if work activity changes

Note 1: http://www.communities.gov.uk/fire/firesafety/firesafetylaw/
Note 2: www.hse.gov.uk/riddor

HSE

Figure 7.11 *A safety policy must be available to all employees*

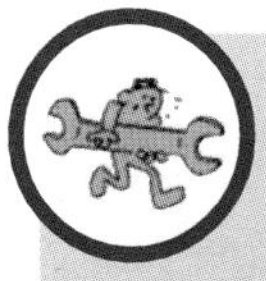

Task

Take a look at the sample safety policy which is available on the HSE website and make a note below of the main points.

____

____

____

____

____

Site visitors

Employers are responsible for ensuring the safety of visitors to their premises or site. Authorized visitors should be asked to identify themselves and the person they wish to see by name. They should then sign in the visitor's book and be made aware of the safety procedures they need to follow while they are on the site.

Remember

Breaking the health and safety laws can result in prosecution.

The health and safety inspectorate

The health and safety inspectorate are the Health and Safety Executive (HSE) and they are concerned with the safety and welfare of everybody at work. They may appoint the Local Authority Environmental Health Department to act on their behalf. Although they have legal powers to prosecute offenders the extent of this depends on the seriousness of the breach. It is far more likely that they will issue either an improvement notice or a prohibition notice. Normally prosecution is only then involved where these notices are not acted upon.

Improvement notice

An improvement notice can be served on an employer if an HSE inspector is satisfied that a statutory regulation has been contravened. Once this notice has been served the employer has a specified time (not less than 21 days) in which to take the required action or to appeal to an Industrial Tribunal if they should wish to do so. During the time that the appeal is pending, the improvement notice is suspended and the equipment may be used. If the employer fails to comply with an improvement notice within the specified time then prosecution will almost certainly follow.

Prohibition notice

A prohibition notice can have the effect of stopping an activity or practice immediately, without recourse to appeal, where the inspector is satisfied that there is a risk of personal injury. Failure to comply with this notice will almost certainly result in prosecution.

Prosecution carries some significant penalties, for example failure to comply with an improvement or prohibition notice can result in a fine of up to £20 000, and/or 6 months' imprisonment. Higher courts may impose unlimited fines and imprisonment.

Remember

An employer must provide:

- Insurance that covers employees in cases of injury or illness due to work
- A safe working environment including facilities for sanitation, washing and first aid
- A safety policy (for workplaces with five employees and over)
- A safe place of work with safe access and exit
- Safe methods of handling and storing goods
- Safe equipment and system of work
- Accident procedure and register
- Training and supervision.

Try this

A maintenance company carried out maintenance to machines and equipment which are used in the motor vehicle repair industry. One item of equipment is an electrically operated side lift hoist used to allow work to be carried out under the car.

1   List the statutory documents which will apply to this maintenance activity.

a   _____

b   _____

c   _____

d   _____

e   _____

f   _____

g   _____

h   _____

2   Who is responsible for producing the safety policy for the maintenance work?

_____

3   An employee has an accident at work and sustains a broken arm.

a   Where must the incident be recorded?

_____

b   Who should be notified?

_____

Part 3

Whilst carrying out maintenance work we need to comply with the health and safety legislation. This will include selecting and confirming the suitability of any personal protective equipment (PPE), tools and equipment that we are going to use.

Personal Protective Equipment

Personal Protective Equipment (PPE) refers to all the equipment intended to be worn or held by a person at work to afford protection against one or more risks to their health or safety.

The Personal Protective Equipment at Work Regulations require that wherever there are risks to health and safety that cannot be adequately controlled by any other means then personal protective equipment must be supplied and used. It is only if there is no other way of providing the protection necessary to do the work safely that PPE should be supplied.

Employers are required to ensure that PPE is:

- Properly assessed before use to ensure that it is suitable for the situation
- Fits the wearer
- Prevents or adequately controls the risks that may occur
- Kept clean, stored and maintained in good repair

- Provided with instructions for its correct use and
- Used in the proper manner by employees.

Remember

If the equipment is provided it must have been assessed as necessary. Never fail to use the PPE provided because the job will only take a few minutes. You also have a duty to ensure that you know, understand and use the PPE equipment correctly.

Footwear

Protective footwear should be worn where there is a risk of injury from manual or mechanical handling, electrical work or any work carried out in adverse environmental conditions.

Footwear should be selected to provide grip, resistance to water or hazardous substances, flexibility and comfort for the wearer. Protective footwear should be worn when there are risks due to heavy materials dropped on the feet and to prevent penetration by nails and other sharp objects.

© Pictures supplied by Draper Tools Limited

Figure 7.12 *Protective footwear*

Eye protection

Safety goggles or spectacles are required in situations such as where there is dust and flying debris, swarf from cutting operations, chemical splashes such as cleansing solvent, airborne debris such as cement dust on construction projects and working overhead.

Figure 7.13 *Eye protection*

Remember

Where eye protection is required in dusty environments dust masks will also be required.

Hard hats

Head protection, more commonly referred to as hard hats, is required where there is a risk of injury from falling or flying objects such as:

● Dangers from fixed protrusions in the work area, such as mounting brackets and pipe clamps

● Material falling from lifting activities such as hoists and cranes

● Material being dropped by persons working above

● Material being kicked into pits and service wells.

Figure 7.14 *Head protection*

Outdoor clothing

When working outside additional outdoor clothing may be required to:

● Offset the effects of wind and rain on the general health of the individual

● Offset particular effects of the elements, long-term exposure to the elements or prolonged outdoor working. These activities may involve work with little or no physical movement such as the termination of cables in a feeder pillar, or extreme physical activity such as digging

● Protect against certain of the elements. This may also be provided by portable shelters, such as the framed work tent used by a cable jointer.

Gloves

Leather gloves can safeguard against cuts resulting from the manual handling of heavy and sharp objects. Gloves may also be required when working outside in very cold conditions.

Hand protection is required:

● Wherever activities may cause risk of skin penetration such as wood or steel splinters
● Where the nature of the material may cause physical skin abrasion such as moving and replacing plant
● Where the materials being handled may be hazardous by either being absorbed into the skin or by transfer to the mouth through contact with food.

Figure 7.15 *Hand protection*

High visibility clothing

High visibility clothing may be required particularly when carrying out maintenance work in certain locations such as:

● Where work is carried out in an area with a high risk from 'passing traffic' such as working on street furniture, street lights and bollards
● In areas with vehicular traffic such as loading bays and docks
● In areas where there are lifting operations being undertaken, such as the construction of high rise buildings.

© Pictures supplied by Draper Tools Limited

Figure 7.16 *High visibility clothing*

Ear defenders

Under the Noise at Work Regulations ear protection should be used in very noisy environments. Maintenance work is often carried out in premises where production and fabrication activities are ongoing. Ear defenders should comply with current BS specifications, fit properly and be comfortable.

A simple guide to determine when ear protection is needed is if:

● You have to shout to be heard by someone who is only 2 metres away

● The noise is intrusive for most of the working day or

● You work with or near sources of very loud noises such as pneumatic drills or power presses.

© Pictures supplied by Draper Tools Limited

Figure 7.17 *Ear protection*

Prolonged exposure to noise can cause temporary and possibly permanent hearing loss which may not be immediately obvious, but develops gradually over time. Another problem associated with exposure to noise is the development of tinnitus (ringing in the ears).

Remember

When working in a noisy environment or wearing ear protection it will be harder for you to hear if people need to warn you of any dangers.

Ear protection is required where there is a high level of background noise, which may not be due to the activities you are involved in.

Try this

List the appropriate personal PPE equipment to be used in the following situations.

1   Carrying out maintenance to low level lighting bollards in a public car park during the normal working day during the summer.

_____

_____

_____

2   Maintaining equipment in a machine shop at the same time as other trades are installing additional service pipes overhead.

_____

_____

_____

Tools and equipment

It is important to use the right tools and equipment when carrying out maintenance work. We have a number of choices to make regarding this equipment and we shall cover the basic requirements here.

Note
There is more detailed information on the hand tools used for many maintenance tasks in the 'Termination and Connection of Conductors' study book in this series.

Hand tools

A general hand tool kit for the electrical maintenance activities will incorporate many common electrician's tools. These will generally include:

- Terminal screwdrivers (small and large)
- Large flat blade screwdriver
- Pozidrive screwdrivers (PZ1 and PZ2)
- Stripping knife (not a Stanley Knife type)
- Electrician's pliers
- Long nose pliers
- Cable strippers
- Side cutters
- Hacksaw
- Junior hacksaw
- Spanners (various sizes)
- Gland pliers (generally two pairs)
- Measuring tape
- Small file
- Crimping tool
- Small pin hammer
- Ball pein hammer.

Roughneck

Figure 7.18 *Ball pein hammer*

© www.rel-Group.com

Figure 7.19 *Gland pliers*

Of course there will be all manner of specialized tools, too many to list here, which may be required depending on the type of work to be undertaken. Common additions can include hand operated crimp tools, a soldering iron and a desoldering gun or braid.

© Courtesy of Draper Tools Ltd

Figure 7.20 *Soldering iron*

On completion of many electrical maintenance activities we will need to carry out tests to confirm compliance. This will require test instruments specifically for this task. These will include:

- Low reading ohmmeter
- Insulation resistance ohmmeter
- Earth fault loop impedance tester
- RCD tester
- Approved voltage indicator
- Multimeter.

Note

Inspection and testing of the electrical installation prior to it being put back in service is covered in detail in the 'Inspection, Testing and Commissioning' study book in this series.

Misuse and mistreatment of tools is a major contributor to accidents in the workplace and one which can be easily avoided. Tools should be checked regularly and damaged tools must be repaired or replaced. Simple precautions can extend the life of tools and regular maintenance improves their operation and longevity.

A few simple rules:

- Only use tools for the purpose they are intended
- Regular maintenance is a must such as a little light oil on the pivot hinges of sidecutters and pliers to keep them moving freely
- Regular checks on the blades of screwdrivers to ensure the blades are square and sharp

- Keep knife blades clean and honed to avoid slips and cuts. Always cut away from you and do not place any part of you in front of the blade
- Make sure hacksaw blades are sharp and inserted correctly.

Power tools

When carrying out maintenance work we often need to use power tools for fixing and removing and installing screwed covers and the like. In some instances we may need to carry out more heavy duty work such as cutting and chasing.

The power tools for many of these activities can be battery operated with the advantage that we do not need a power supply and there are no trailing leads and the like to contend with or create trip hazards. Having two batteries for these tools means that one can be charging whilst the other is in use and this will therefore extend the working period for the tool indefinitely (assuming a power supply is available for recharging). Battery operated lighting units, also rechargeable, may be used for areas where additional localized lighting is required.

For heavy duty applications we may need heavier duty power tools supplied from the electricity supply. The preferred voltage for these power tools is 110 V supplied through a reduced low voltage transformer. This transformer is a centre tapped to earth, step-down transformer which limits the line voltage to 55 V above earth potential.

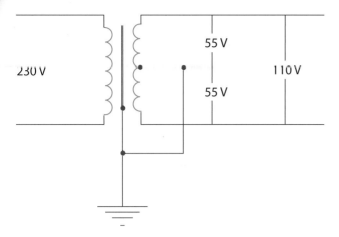

Figure 7.21 *110 V centre tapped to earth*

The flexible cable used for these 110 V tools and equipment is coloured yellow and connections are by way of BS EN 60309 plugs and sockets. Where 230 V power tools have to be used these must be protected by an RCD rated no more than 30 mA and the flexible cable is coloured blue, again using BS EN 60309 plugs and sockets. The pin configuration for these plugs and sockets is different and so prevents 110 V equipment from being plugged into 230 V sockets and vice-versa.

Where mains supplied portable electrical equipment is used in the workplace accidents can often be prevented by following a few simple rules. A visual check on cables and plugs, which are particularly liable to damage, can prevent a serious accident. Electrical equipment may develop faults which do not affect the operation but may present a potential hazard. Inspection and regular testing of this equipment and implementing any repairs helps to ensure accidents are prevented.

Note

More information on the BS EN 60309 plugs and sockets and the checking of portable equipment is given in the 'Legislation Health, Safety and the Environment' study book in this series.

Lifting equipment

There will be occasions when our maintenance work involves the replacement or installation of large items of equipment. At other times the materials required, such as lamp replacement in a large office or workshop, may be delivered in bulk packaging. In either event we will need to move the materials to an appropriate location and may need to lift them into position.

Moving the materials may be achieved in a number of ways:

- Small quantities or items may be carried to their location and lifted into position
- Larger quantities may be moved using a pallet truck, sack barrow or flat truck depending on the material involved
- Large quantities may be moved using a fork-lift truck or similar device.

Once the materials or equipment are at the worksite we may need to use lifting gear to get it into position. Depending on the equipment and the lift required we can use a number of different lifting aids including:

- Simple block and tackle to lift materials into position, often used to lift material onto scaffolding etc.
- A power hoist can be used in the same way as the block and tackle. They are often mounted on a mobile base for lifting and moving motors and similar equipment into their location
- A fork-lift truck may be used to lift materials onto working platforms and scaffolding
- Cranes may be required for serious lifting activities such as replacing an air conditioning plant on the roof of a building

- Hydraulic or scissor jacks may be used for lifting heavy equipment short distances to their final mounting height.

Figure 7.22 *Mobile hoist*

Equipment for adjustments

In addition to the lifting equipment we may need to use equipment to position, move and reform materials and equipment. We have seen that some of the lifting equipment may be used to assist in the positioning of equipment but often small adjustments are necessary to achieve correct positioning or alignment.

This type of equipment will include:

- Rams which are a form of hydraulic jack can be used for both lifting and straightening by applying the pressure internally or externally to damaged or misaligned components. They can also be useful for moving heavy equipment small distances for positioning

- Straightening and rotating devices are used to adjust cables, pipework ducting and other similar services. They allow small movement and repositioning to ensure alignment. Large

rotating devices may be required to change the orientation of large items of plant say from the horizontal to vertical mounting position

- Fin straighteners may be used when maintaining equipment with aluminium cooling fins as these are easily bent and damaged. This will affect their performance and so special tools are used to straighten the fins.

Figure 7.23 *Typical adjusting tools*

Figure 7.24 *Fin straightener*

Task

Compile a list of the hand tools that you would need to carry out the maintenance on a fixed warm air heater in a workshop.

_____

_____

_____

_____

_____

_____

_____

_____

_____

_____

_____

_____

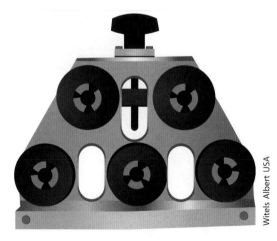

Figure 7.25 *Wire/cable straightener*

Witels Albert USA

Note

Further information on lifting and handling and access equipment can be found in the 'Legislation, Safety and the Environment' study book.

Access equipment

Maintenance work often involves the use of access equipment to get to the actual worksite. This may be a simple pair of steps for short duration work such as changing a lamp or a fixed scaffold.

for long duration work at height such as the maintenance of air conditioning units mounted at high level on the outside of an office block.

It is important to select the appropriate access equipment which could include:

- Steps and ladders for short duration work such as lamp changes
- A step up for working at lower levels for longer durations such as replacing a luminaire
- Platforms and trestles for working over larger areas
- Tower scaffolds for work at higher levels such as the maintenance of high bay lighting. These are often mobile systems which can be moved from one location to another
- Mobile elevated work platforms (MEWP) for work on high level outside lighting and equipment.

Remember

Whenever access equipment is to be used the manufacturer's instructions should always be followed.

We can see that different access equipment will be required depending on the height at which work is to be carried out and the time the work will take. Reaching just above shoulder height may only require a step up, whereas scaffolding may be required for access to work on the roof of a building.

There are some basic rules which apply whichever access equipment is required:

- All access equipment should be set up on a firm level base

- The equipment chosen must be suitable for the task so that the user does not have to over-reach
- All access equipment should be inspected regularly to ensure that it is in good condition. This does not just mean whether or not it is broken, it also includes checks like looking to see if the surface is slippery because of mud or ice and other relevant hazards.

Note

Further information on the use of ladders and stepladders can be found in HSE Guidance available as free downloads from the HSE website www.hse.gov.uk.

Materials

We have considered the requirements for the materials earlier in this study book and so here we shall consider the properties of the materials which will help in making the correct selection.

Thermoplastic: the most common form of this type of material is PVC used for the insulation of conductors. Due to its structure it does not have mechanical strength and therefore needs to be protected either by a sheath or by installation within a containment system or equipment. The structure of PVC changes at around 80 °C and it becomes softer and pliable and so thermoplastic cables have a maximum conductor operating temperature of 70 °C

Thermosetting: the most common form of this type of material is also for the insulation of conductors; cross-linked polyethylene (XLPE) being the most common. The structure of this material is such that it is less affected by temperature and so these cables have a maximum conductor operating temperature of 90 °C.

Fibreglass: used in the form of woven strands this is often used as sleeving for conductors and insulation in locations where high temperatures may be experienced. These sleeves are often used in the terminal compartments of floodlights and similar fittings. It is also a common insulation material for the conductors within electric ovens, hobs and other locations where high temperatures may be present.

Sleeving: this may be made from any of the insulating materials and is commonly used to identify or provide addition protection to the insulation of conductors. Colour coding of conductors and protection of cpcs at terminations for flat twin profile cables are common uses.

Shrink sleeving is another variant of sleeving which is often used during maintenance. The most common is heat shrink sleeving, although cold shrink sleeving is available. There are different types available for different uses. The electrical shrink sleeving is used to provide protection to terminals and terminations, provide additional protection for insulation and abrasion resistance and to group cables together in equipment. It may also be used in the repair of damaged insulation providing a suitable grade and process to be used.

Heat shrink: is a mechanically expanded tube which when heated returns to its original size and it is rated by its expansion ratio. So for example it may reduce to a quarter of its expanded size when heated and providing the host is between the expanded size and not less than a quarter of the expanded size, when heated the sleeving will shrink to form a tight seal onto the host.

Varnishes: synthetic varnishes are used as insulators and protective coatings in control equipment and motor windings. It is mechanically strong and durable and provides good electrical insulation for a very thin coating. It is applied in several thin layers to obtain the final insulation but this is thinner than other types of insulation. Shellac is an earlier natural resin which is secreted by the female Lac bug. This is processed, dried and then mixed with ethyl alcohol to produce the shellac. This has largely been replaced by the synthetic alternative lacquers and varnishes.

Ceramics: ceramics are used for insulators in some equipment and distribution and transmission systems. They are basically glazed clay which is fired to seal it and stop it from being porous. They are used extensively for the insulators for overhead supply cables as the insulator must not only provide insulation but have sufficient mechanical strength to support the conductors. Ceramic mounts are also used within busbar chambers to insulate and support the bars away from the metallic enclosure, although these are being outmoded by modern plastic insulators.

Metals: the considerations for the metals used in the installation, equipment and the building structure were discussed earlier in this study book. We need to ensure that electrolytic reactions do not take place between dissimilar metals. One common method of overcoming eddy currents at terminations is to use an aluminium gland plate in place of a steel one. If armoured cables are to be terminated in this gland plate the glands must not be made of brass.

Solvents: are used in many maintenance applications from cleaning to sealing and jointing. In addition to the COSHH requirements many solvents have detrimental effects on other materials. The solvent adhesive used for jointing PVC conduit does so by reacting with and softening the material and fusing it together. If it is spilt on other plastic or polished surfaces the surfaces will be damaged. Solvents used for cleansing prior to jointing and those used as part of electrical jointing processes can all have a damaging effect on other materials. It is important to be familiar with the manufacturer's instruction and COSHH information when selecting and using solvents.

 Try this

Identify the most appropriate access equipment for each of the following tasks.

1   Replacing a discharge lamp in a luminaire mounted at a height of 4 metres on an outside wall.

_____

2   Replacing the control gear in a luminaire mounted on a 5 metre lighting column.

_____

3   Maintaining an air conditioning unit which is ceiling mounted at a height of 3 metres.

_____

 Try this

Select the most appropriate material for each of the following:

1   protecting the insulation of conductors at their termination in a 500 W tungsten halogen floodlight.

_____

2   insulation of conductors with a normal operating temperature of 70 °C.

_____

3   supporting bare overhead conductors mounted on a pole with metal cross arms.

_____

4   jointing PVC conduit and trunking.

_____

Part 4 Inappropriate work practices

Preventing accidents

We considered the requirements of statutory health and safety legislation at the beginning of this chapter. A brief recap on the employee's responsibilities is worthwhile before we look at inappropriate working practices.

Employees are required by law:

- To take reasonable care for their own health and safety and not to endanger others
- To cooperate with their employer on health and safety procedures
- To not interfere with tools, equipment, etc. provided for their health, safety and welfare
- To correctly use all work items provided in accordance with instructions and training given to them.

Remember

An employee can be prosecuted for breaking the health and safety laws.

Accidents at work

An accident is an unexpected or unintentional event which is often harmful and accidents at work do not just happen they are caused by people.

The main causes of accidents are people who have:

- Become unsafe because of such factors as:
  - Boredom
  - Horseplay
  - Carelessness

- Taking shortcuts
- Lack of knowledge.
- Been provided with or produced an unsafe environment
- Misused a safe environment.

Inappropriate actions or lack of action can result in injury or death to us or other people.

Common situations which occur during maintenance and their potential consequences are:

Failure to isolate circuits and equipment: the implications for this were discussed in Chapter 5 of this study book. The reason for this failure is often because the operative believes this will save the time taken to carry out the safe isolation procedure. This can be prompted by the desire to finish the work quicker either as the result of pressure from the client or those using the equipment being maintained. This seldom saves any time and the risk of electrocution to us and other people is severe. Working on or near live equipment should be avoided at all times and EWR identifies that where this must be carried out then suitable procedures and equipment must be used. The net results are the hugely increased risk of electric shock, arcing and electrical burns. Once such an event occurs the protective device should operate and will probably result in even greater disruption and loss of supply. The failure to isolate would be considered as a breach of EWR and could result in prosecution of the operative.

Bypassing safety features and interlocks: safety features and interlocks are installed to prevent danger and by their nature make sure that the supply can only be energized when the correct procedure is followed. When maintaining the equipment or installation in order to check the operation operatives may bypass the safety features. This places the operative and those in the vicinity at risk of electric shock, burns and injury from the equipment operation.

Not obtaining permission to isolate or switch off: when working in occupied and functioning installations we must always seek permission before isolating supplies or switching off equipment. By not obtaining permission we run the risk of causing damage to equipment, interrupting processes or losing important data. All of these can have significant cost implications for the client's business and in some instances this could run to millions of pounds particularly where lost data are involved.

Failure to use barriers and signs: when we are working in occupied premises it is essential to ensure that people using the premises are not put in danger. The failure to use barriers to keep people away from our work area when carrying out maintenance places both them and us in danger. We must also ensure that any open access pits, switch rooms and locations where we are working overhead are protected by barriers to prevent injury from falls, electrocution and material falling from above.

Posting notices: to warn that work is being carried out advizing people as they enter the work area will make people aware, and remind them, of the work that is being carried out. Failure to do so means that people will not be expecting anything unusual and not taking particular care; possibly placing them in danger.

Use of inappropriate or unsafe access equipment: there are many occasions where the wrong type of access equipment is used for expediency, because it's here it's easier than get-

ting the right one. So carrying out work using make-do or botched access equipment can result in falls and injuries. Incorrect use of the correct access equipment can also cause problems such as back, muscle and joint injuries and these, whilst not always immediately noticeable, can be painful and in severe cases prevent you from working.

Figure 7.26 *Unsuitable access equipment*

The use of additional safety equipment when working at height is essential. Fall arrestors, safety harnesses and the like are often not used because they are inconvenient or don't look smart. These are essential for your safety and failure to use them can result in serious injury or death. For example whilst working in MEWP to maintain street lighting it requires only a small bump from a passing vehicle to cause violent movement of the platform. Without a safety harness the operative can be thrown off the working platform and suffer severe or fatal injuries as a result.

When using the appropriate access equipment it must be used in the correct manner. For example the incorrect use of a pair of steps to drill a fixing hole can result in a fall and the resulting injury. When using ladders and steps work should always be carried out without the need to reach out to the sides. Failing to do so may result in falls and injury.

Figure 7.27 *Use steps and ladders correctly*

Misuse of tools or equipment: tools and equipment are made to perform particular functions and using them for other types of work can result in both damage to the tools and injury to the operator. Using a screwdriver as a chisel, for example, can damage the blade and weaken the handle. The handle can then fail at a later date resulting in injury to the user. Levers and crowbars are designed for opening crates and lifting flooring etc. and using screwdrivers for this job will again result in damage to the tools and possible injury to the user.

Failing to maintain tools and equipment in good condition: it is important to maintain the tools of our trade in good condition as this affects their efficiency and our safety. We have considered the need to keep pliers and the like lubricated and operating smoothly. We need to check that the shaft of hammers are not damaged or split as the shaft breaking during use can result in serious injury or damage to the building fabric and equipment. The heads of chisels and bolsters need to be kept free of burrs or 'mushrooms' as they are often referred to. Failing to do so results in these breaking and flying off during use resulting in cuts and bruises, and without proper eye protection can result in the permanent loss of eyesight.

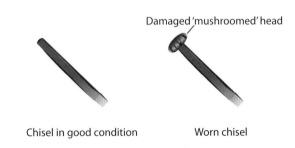

Damaged 'mushroomed' head

Chisel in good condition   Worn chisel

Figure 7.28 *Maintain chisels in good condition*

We can see from the above that even when the right equipment is provided misuse by the operative can result in injury to themselves and others. Failure to follow correct procedures may result in injury and loss of data or production. It is important that the correct procedures, equipment, materials and tools are used and used correctly at all times.

Remember

Accidents don't just happen – they are *caused*.

You can help to minimize the number of accidents that occur by:

● Learning the safety rules of your workplace and industry and following them

● Protecting yourself and the people around you by using the correct equipment, clothing and gear

● Keeping a look-out for potential hazards and reporting those you find to the appropriate responsible person

● Behaving sensibly and thoughtfully.

Try this

List one possible outcome for each of the following.

1 Not obtaining permission to isolate the supply to a circuit supplying computer equipment.

_____

_____

2 Not isolating the circuit when carrying out work on the control panel for a processing machine.

_____

_____

3 Failing to provide barriers around the cover which has been removed to access the underfloor sockets supplying office equipment.

_____

_____

4 Using a screwdriver and hammer to remove knockouts from a replacement enclosure.

_____

_____

Congratulations you have now completed Chapter 7 of this study book. Complete the self assessment questions before you continue to the next chapter.

1   Anything with the potential to cause harm is described as a:

   a.   Risk

   b.   Hazard

   c.   Danger

   d.   Failure

2   Following risk assessments a company must produce a:

   a.   Contract

   b.   Disclaimer

   c.   Safety policy

   d.   Working manual

3   PPE should be issued when:

   a.   The company can afford it

   b.   The employee requests it

   c.   When the risk cannot be removed

   d.   When the HSE states it must be provided

4   The preferred voltage for mains operated power tools used for maintenance is:

   a.   230 V ac

   b.   110 V dc

   c.   55 V ac

   d.   55 V to earth

5   One of the main causes of accidents when carrying out maintenance work is not:

   a.   Using the correct tools

   b.   Taking shortcuts

   c.   Being careless

   d.   Being bored

Principles and regulatory requirements for completing worksites for maintenance

8

RECAP

Before you start work on this chapter, complete the exercise below to ensure that you remember what you learned earlier.

A risk _____ must identify the _____ taken to _____ the risks and these actions must be _____ out.

A hazard is anything with the _____ to cause _____ and a risk is the _____ of a hazard causing _____ or damage.

The provision of personal protective equipment should _____ be considered as a _____ resort when there are no other _____ measures which can be_____.

A safe working environment includes such factors as a clean and _____ workplace and _____ and _____ lighting.

Dangerous _____, major injuries and _____ must be reported to the relevant _____ authority.

Major injuries include those that result in injury from _____ shock requiring _____, fractured bones, loss of _____ or hospitalization for more than _____.

A HSE _____ notice can have the effect of _____ an activity or practice _____, without recourse to _____ .

It is important to use the right tools and _____ when carrying out maintenance work and, on completion of electrical maintenance activities, _____ are carried out to confirm _____ .

Where power tools are supplied from the mains supply the preferred voltage for these power tools is _____ supplied through a _____ low voltage _____ .

During maintenance equipment may be required to_____ materials or _____ into position and to move or _____ materials or equipment.

It is important to _____ the appropriate access _____ for the type and _____ of the work to be undertaken.

An accident is an unexpected or _____ event which is often _____ and they are caused by _____ .

Tools and equipment are made to perform particular _____ and using them for other types of _____ can result both in _____ to the tools and _____ to the operator.

LEARNING OBJECTIVES

On completion of this chapter you should be able to:

- Identify and interpret appropriate sources of information relevant to maintenance activities including:

  - Statutory documents

  - Codes of Practice

  - British Standards (including current version of BS 7671)

  - Maintenance schedules

  - Manufacturer's guidance documents.

- Interpret diagrams, drawings, maintenance schedules and specifications to identify the replacement/refitting requirements of wiring systems and equipment as applicable to maintenance procedures including:

- Cables

  - Thermosetting insulated cables including flexes

  - Single and multicore thermoplastic (PVC) and thermosetting insulated cables

  - PVC/PVC flat profile cable

  - Mineral-insulated copper-clad (MICC) (with and without PVC sheath)

  - Steel wire armoured (SWA) cables (paper insulated lead covered (PILC), XLPE, PVC)

  - Armoured/braided flexible cables and cords

  - Data cables

  - Fibre optic cable

  - Fire resistant cable.

- Equipment

  - Electrical plant, components and accessories

  - Motors and starters

  - Switchgear and distribution panels

  - Control systems and components

  - Contactors

  - Power transmission mechanisms

  - Luminaires and lamps

  - Drive systems.

- State the work methods and procedures for completing maintenance activities for particular systems and equipment

- State the requirements for completing appropriate corrective actions/repairs when problems are identified, including:

  - When it is appropriate to carry out repairs

  - The advantages and limitations of repair against component replacement

  - Responsibilities for making decisions regarding repairs that are required

  - Approved procedures for the completion of repairs

  - Likely implications for relevant parties of carrying out effective repairs.

As you work through this chapter you may find it useful to refer to the statutory and non-statutory publications referred to and these may be either as hard copy or, in many cases, accessed on-line.

Part 1

We have considered the statutory and non-statutory documents relating to maintenance earlier in this study book. In Part 1 of this chapter we are going to consider the requirements of the documentation and regulations that are relevant to carrying out electrotechnical maintenance work. These include:

- Health and Safety at Work (etc.) Act
- Provision and Use of Work Equipment Regulations
- Electricity at Work Regulations 1989
- Memorandum of Guidance on the Electricity at Work Regulations 1989
- BS 7671, Requirements for Electrical Installations.
- IET Electrical Maintenance
- Code of Practice for the In-service Inspection and Testing of Electrical Equipment
- Maintenance records
- Manufacturer's information

Health and Safety at Work (etc.) Act

The Health and Safety at Work (etc.) Act applies to everyone who is at work. The aim is to:

- Improve or maintain the standards of health, safety and welfare of all those at work
- Protect others against the risks arising from work activities
- Control the use and storage of hazardous substances.

The requirements will affect our preparations, actions and precautions during the maintenance work.

PUWER (Provision and Use of Work Equipment Regulations)

PUWER relates to all equipment used at work and generally, any equipment used by an employee during their work is included. Equipment supplied by an employer must be suitable for its intended use and be safe for use.

In addition to the employer's responsibility we, the users, have a responsibility to ensure the equipment is safe before we use it.

Figure 8.1 *HSE Guidance 'Safe Use of Work Equipment'*

The Regulations state that any risks that are created by the use of the equipment should be controlled by such means as protection devices, guards, warning devices and personal protective equipment (PPE). We have a responsibility to confirm these safety measures are in place and provide the necessary protection for the task being undertaken.

Electricity at Work Regulations 1989

The Electricity at Work Regulations (EWR) are statutory and are particularly relevant to electrical maintenance. Every employer, self-employed person and employee is required to comply with safe working procedures to ensure electrical safety in the workplace.

When undertaking maintenance we are required to:

- Take reasonable care for our health and safety
- Not endanger others
- Cooperate with our employer on health and safety procedures
- Not interfere with tools, equipment etc. provided for our health, safety and welfare
- Correctly use all work items provided in accordance with instructions and training received.

The Memorandum of Guidance on the Electricity at Work Regulations

The memorandum of guidance provides information and guidance on the requirements of the EWR and how this can be achieved. It

Figure 8.2 *The Memorandum of Guidance on the Electricity at Work Regulations 1989.*

includes useful information which helps us to interpret the requirements of the statutory document.

The normal practice required under the Electricity at Work Regulations is to work with the circuit or equipment isolated from the supply. We should ensure that, other than when carrying out live testing, that any circuit or equipment we are working on is safely isolated. We must also ensure that a method of preventing accidental reconnection, such as locking off, should be used to protect us whilst we are carrying out maintenance.

Note

The procedures and requirements for safe isolation are covered in Chapter 6 of this study book.

BS 7671

BS 7671, the Requirements for Electrical Installations, is published by the Institution of Engineering and Technology (IET) and is commonly known as 'the Wiring Regulations'. BS 7671 provides the requirements for electrical installations and any maintenance work we carry out on electrical installations must comply with BS 7671. Any such maintenance work must be compliant with the current requirements and the existing installation must be no less safe. BS 7671 provides guidance on the requirements for the installation and the inspection and testing of the work before placing it back into service.

IET Electrical Maintenance

The IET Guidance on Electrical Maintenance provides guidance and information on the requirements for electrical maintenance of a number of types of system and equipment. The topics include;

- Electrical installations
- Testing
- Lighting maintenance
- In-service inspection and testing of electrical equipment
- Emergency lighting
- Fire detection and alarm systems
- Industrial and commercial switchgear
- Electromagnetic compatibility
- Lightning protection
- Environmental protection
- Legionellosis.

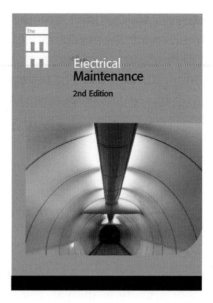

Figure 8.3 *IET Electrical Maintenance*

Code of Practice for the In-service Inspection and Testing of Electrical Equipment

This code of practice (CoP) includes the requirements for the inspection and testing of electrical equipment provided for use in the workplace and public access areas. Part two and the Appendices of the CoP provide information and guidance for those persons responsible for the inspection and testing of the electrical equipment. This includes details of the tests and expected results. It also contains information and guidance for the duty holders responsible for the maintenance.

Maintenance schedules

Maintenance schedules provide us with the details of the work which has been carried out and what work is required for planned preventative maintenance. This information is in addition to that given in the manufacturer's information as the client will have their own maintenance programme. This will have been developed with consideration to the manufacturer's recommendations and may be more onerous due to the client's particular activities or requirements.

Manufacturer's information

The manufacturer's information usually contains information which is necessary for the correct maintenance and operation of equipment including:

- Technical information such as power requirements, current and voltage ratings
- Wiring and layout diagrams
- Setting and adjustment information for correct operation
- Maintenance requirements
- Fault finding charts
- Spare parts information and references.

Figure 8.4 *Typical manufacturer's instructions for small appliance*

© Consort Equipment Products Ltd.

Task

Maintenance is to be carried out on an air conditioning unit located in the ceiling of an office reception area. Identify:

1 **Two statutory documents which are relevant to the maintenance activity.**

_____

_____

_____

2   Two reference documents which may be referred to during the maintenance of the air conditioning unit.

_____

_____

_____

3   Two areas of consideration relating to the access to the unit to undertake the work.

_____

_____

_____

4   Two considerations relating to the safety of other persons using the office building whilst the maintenance work is undertaken.

_____

_____

_____

Part 2

The study book 'Installing Wiring Systems' in this series identifies the types of wiring systems and some of their uses. In this chapter we will consider the wiring systems for typical applications in maintaining electrical installations. There will be a need for you to refer to manufacturers' information, BS 7671 and IET Guidance Note 1, when you are working through this chapter. You may also find the IET On-Site Guide helpful.

We have considered the selection of components, materials and equipment based upon their properties and the requirements for performance earlier in this study book. In addition we have to consider the physical requirements and specification which will include the use of diagrams, drawings, schedules and specifications.

When determining the requirements for the maintenance we may need to refer to diagrams to establish the requirements. It is often difficult to investigate and measure requirements for maintenance on-site due to the normal operation of the business. In such cases we can use the reference material to determine what is required.

The type of installation system and cables will need to be established and on some occasions this may be quite obvious. For example the majority of wiring in dwellings is in flat profile multicore cable. This flat profile cable is also used for a number of commercial and industrial locations. It is a very versatile wiring system, and with suitable additional mechanical protection where appropriate, can be used in a wide variety of installations.

However this does not automatically mean flat multicore cable is a suitable system and a combination of the use of the premises and the environmental conditions must always be considered. The cable specification may be less apparent as there may be a need for thermosetting cables in some types of installation and low-smoke and fume (LSF) and fire rated cables for commercial properties and houses of multiple occupation.

The types of cables and methods of installation are varied and we will consider the installations in terms of their use for lighting and power.

We will begin by looking at the requirements for lighting systems. There are some common installation systems used for lighting and the choice is normally based upon a number of factors:

● The extent and complexity of the lighting system
● The nature of the building
● The intended use of the building
● The requirements of the client.

Before considering the maintenance requirements for lighting circuits we should first consider the basic lighting circuits. For this exercise we shall use the example of an installation carried out in flat multicore and circuit protective conductor (cpc) cables.

Basic lighting circuits

The most common domestic lighting circuit is generally a three-plate wiring system which may be carried out using a suitable junction box. The circuit diagram for this is shown in Figure 8.5 and the practical circuit in Figure 8.6.

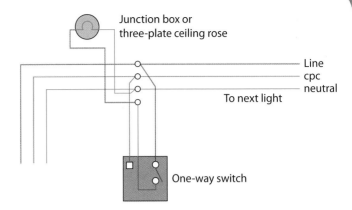

Figure 8.5 *Three-plate lighting circuit diagram*

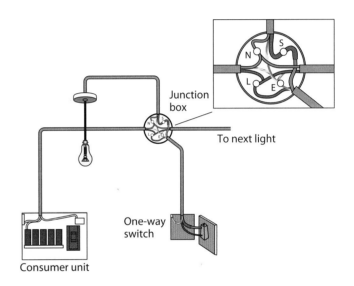

Figure 8.6 *Three-plate junction box*

> **Note**
>
> BS 7671 requires all such terminations in equipment to be accessible unless the equipment is identified as complying with BS 5733 and marked as 'maintenance free' with the symbol MF and installed in accordance with the manufacturer's instructions.

Three-plate ceiling roses are commonly used in the installations in dwellings and plug-in three-plate versions are widely used in commercial and industrial installations.

Figure 8.7 *Typical plug in ceiling rose*

The linking of the conductors may also be carried out at the switches providing the connections are contained within suitable terminations.

Two-way switching

Sometimes it is necessary to have two switches controlling the same light(s). This often happens on staircases or locations where access is afforded in two locations and in these situations two-way switches are used. When using flat multicore and cpc cables this circuit is usually constructed using three-core and cpc cable. This is known as the conversion method, as by the use of three-core and cpc cable and a change of switch a one-way light may be readily converted to a two-way.

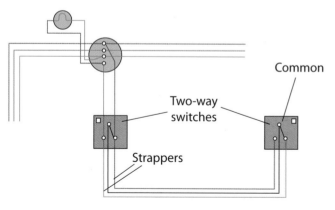

Figure 8.8 *Two-way lighting circuit diagram*

Intermediate switching

Where more than two switching positions are required a two-way and intermediate circuit is used. The intermediate switch has four terminals and is used for all switches between the two-way switches fitted at either end of the switch line. The internal switch connections are as shown in Figure 8.10.

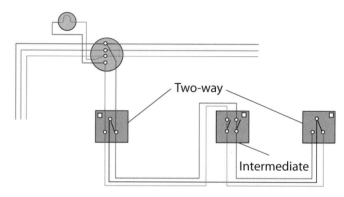

Figure 8.9 *Circuit diagram for a two-way circuit with one intermediate switch*

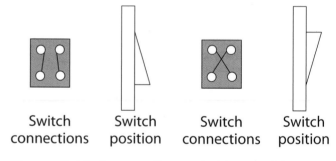

Switch connections   Switch position   Switch connections   Switch position

Figure 8.10 *Intermediate switching positions.*

Note

Where cable cores coloured other than brown are used as switch wires they should be correctly colour coded using brown sleeving or tape to indicate that they are line conductors.

These basic circuits can be modified by the inclusion of presence sensors, photocells, passive infra-red detectors and energy saving measures such as daylight reduction. The basic circuits remain the same.

Having considered the basic requirements for the operation and switching of basic lighting circuits we need to identify the installation system. The common systems available include:

- Flat multicore and cpc cable
- Lighting trunking
- SWA cables
- Conduit and trunking
- Mineral insulated metal sheath cables (MIMS) and FP cables
- Modular wiring systems.

The type of wiring system and lighting controls used will dictate the requirements for any spare or replacement parts required during the maintenance of the system.

Flat multicore and cpc cables: are suitable in locations where they are not subject to adverse conditions or mechanical damage. Additional protection may be installed to enable these cables to be used in some more onerous locations. With suitable precautions they may be used in commercial installations, such as above suspended ceilings, but adequate support such as cable tray needs to be provided.

Lighting trunking: is ideally suited to locations where long runs of lighting are to be installed and the containment system for the cables can be readily installed and may be surface mounted. This is predominantly found in commercial and industrial locations where such installation methods are both possible and acceptable. This system provides a dedicated containment for the lighting system and support for the luminaires. It is robust and versatile and allows the use of single-core cables.

Copyright Legrand Electric Limited

Figure 8.11 *Typical lighting trunking*

SWA cables: are often used for locations where a dedicated circuit is to be installed and the system is unlikely to need alteration or modification. For example high-bay lighting is often carried out using this cable, with fixings direct into the structure or attached to steel girders and the like. The advantage of the good mechanical protection provided is often the reason for selection particularly in locations where the risk of mechanical damage is quite high.

Conduit and trunking: are often used in industrial locations and may be used in larger commercial buildings. These have the advantage of providing mechanical protection for single-core cables, shared routes for other circuits such as power and socket outlets and versatility for the installation of additional circuits or rewiring existing circuits. This system is often difficult to conceal as the containment system does need to be accessible for drawing in cables.

Mineral insulated metal sheath cables (MIMS): offer excellent fire resistance properties, are robust and compact in size compared to other cable systems and when correctly installed outlast most other types of wiring systems. They are ideally suited for locations of national, commercial, industrial or public significance, such as national museums and cathedrals. MIMS may be used for life protection systems such as emergency lighting, where the equipment is to continue to perform under fire conditions.

FP cables: are lower cost alternatives which have fire performance rated insulation and internal aluminium sheathing and the construction and material used allows the cable to continue to perform under fire conditions. These cables will continue to perform providing they are not disturbed under fire conditions. This proviso requires the cable to be suitably supported throughout with fire rated cleats or saddles.

Modular wiring systems: are used in many commercial office buildings where they provide a flexible system with the facility to provide sophisticated control systems. As a plug and play system it provides a great deal of flexibility and adaptability for the user. Savings for commercial concerns, in terms of power consumption and lowering carbon footprint, can make these systems cost-effective in a relatively short period. The system may require additional support such as cable tray to prevent damage to the interconnecting cables.

Distribution board
The male connector on the Master Distribution Box (MDB) being connected into the female socket on the distribution board.

The MDB has been installed on the wire cable tray, the home run cable has been run back to the distribution board. Extender leads are being plugged in to provide a supply to power and lighting loads.

Figure 8.12 *Typical modular system*

Try this

A two-way and intermediate lighting circuit is installed in a steel conduit system using single-core cables. Draw a simple circuit diagram for this lighting circuit where there will only be two strappers between the two-way and intermediate switches. The switch line will go to one two-way common terminal and the line to the other two-way common terminal.

Part 3 Power systems

Power circuits include both electrical equipment and socket outlet circuits and the most common power circuit systems include:

- Skirting trunking
- Dado trunking
- Bench trunking
- Modular wiring systems
- Powertrack and busbar trunking
- SWA cables
- Conduit and trunking
- MIMS and FP cables
- Flat multicore and cpc cable.

The three varieties of trunking, skirting, dado and bench, all provide a containment and mounting system for accessories. Whilst they may be used for multicore cables they are generally used with single core cables. As the names suggest each trunking is designed for use in particular locations. Their use is primarily associated with the installation of socket outlets for power and data and to provide a containment system for fixed equipment wiring. In many cases the trunking is segregated (contains more than one compartment) which allows different systems to be installed within the same trunking system.

The common application for these containment systems are in commercial and industrial installations. The skirting and dado trunking are commonly used in office locations where the system selected depends on the use of the installation. Executive style offices often include skirting trunking whilst in more general office areas the access to accessories may require dado trunking. The bench trunking is generally used in workshop and laboratory locations to allow ready access to accessories.

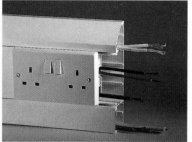

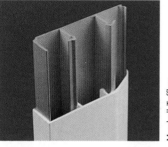
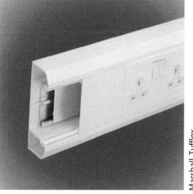

Figure 8.13 *Skirting trunking, dado trunking and bench trunking*

Powertrack and busbar systems: are basically a prewired distribution trunking which uses copper bar conductors and plug-in units to tap off power supplies where needed. These systems are commonly used in industrial locations to provide a versatile distribution system for plant and machinery. The tap off units incorporating protective devices allow for a number of types of equipment with different power requirements to be supplied via a single power distribution system. The powertrack system is frequently used below raised floors in commercial premises to supply socket outlets and equipment. The use of bar conductors means there is no wiring to be carried out up to the tap off points.

SWA cables: are frequently used to supply individual items of equipment in industrial locations. The fact that they are robust and readily fixed to a variety of surfaces, installed in ducts or directly in the ground makes them a very versatile option. These are sometimes used as an alternative to conduit and trunking systems in smaller industrial locations.

Conduit and trunking: is a common installation method in commercial and industrial installations. It provides a containment system for single core cables and can be used for a number of circuits sharing a common route. The use of plastic conduit and trunking in areas where the risk of mechanical damage is not high can prove economical in terms of installation time and material with fewer specialist tools required. Steel conduit will provide a robust containment system suitable for industrial locations, and this may be used in commercial premises, warehouses and the like.

MIMS or FP cable systems: are used to supply power to safety services such as sprinkler pumps, firefighters' lifts, smoke vents and alarm systems. These are selected for their performance under fire conditions and the equipment associated with safety services must continue to function in the event of a fire. They may be used to supply other items of equipment but this is not common.

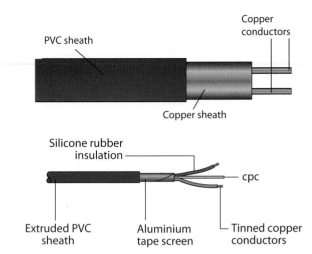

Figure 8.14 *MIMS and FP cables*

FP200 Fire Resistant Cable

Flat multicore and cpc cables: are one of the most common methods used to supply socket outlet circuits in dwellings and commercial properties. The advantages and applications are similar to those discussed for lighting systems.

Socket outlet circuits

The ring final circuit

The most common socket outlet circuit used in domestic installations is the ring final circuit. As the name implies, the circuit starts and finishes at the same point and should form a continuous ring. The line starts and finishes at the same protective device, the neutral starts and finishes at the same connection on the neutral bar, and the circuit protective conductor starts and finishes at the same connection on the earth block. Each socket outlet on the ring must have at least two conductors in each terminal.

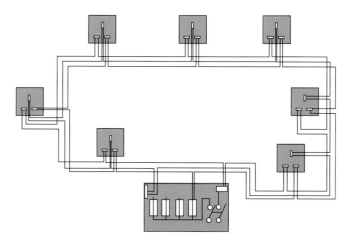

Figure 8.15 *A ring final circuit*

Non-fused spurs

Where a socket is required at a point away from the run of the ring circuit cables a non-fused spur may be used. This is a single cable run just to this outlet. The cable conductors for a non-fused spur must not be of a smaller cross-sectional area than that of the ring conductors.

The non-fused spur may be connected from:

● A socket on the ring
● A junction box connected onto the ring or
● The consumer unit or distribution board.

A non-fused spur must not supply more than one single or one twin socket outlet or one fixed appliance connection unit.

Radial circuits

Radial circuits are basically a circuit with conductors which loop in and out of each item of equipment or accessory.

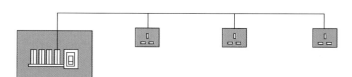

Figure 8.16 *Radial socket outlet circuit*

Distribution circuits

Distribution circuits (often referred to as sub-mains) are used to supply distribution boards which are remote from the origin of the installation. Distribution circuits are common in complex installations where there are a number of distribution boards supplied from a single incoming supply.

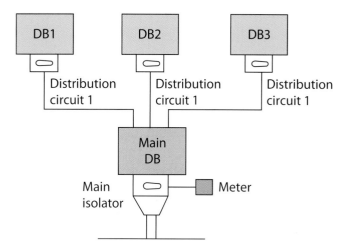

Figure 8.17 *Typical distribution circuit configuration*

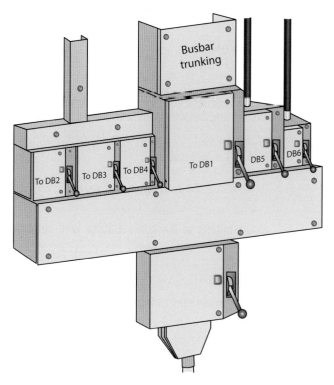

Figure 8.18 *Typical distribution layout using a variety of systems*

The principal systems used for these distribution circuits are:

- Busbar trunking
- SWA cables
- MIMS.

The system used will depend upon the purpose of the distribution circuit, the construction of the premises and load.

Busbar trunking: is often used in multi-storey buildings where a large rising busbar is installed vertically through the building. At each floor there will be tap off boxes supplying distribution boards and then the final circuits. Whilst this system is effective and efficient the busbar is not flexible and whilst it can be supplied with bends and sets it is not suited to installation in restricted access areas.

SWA cables: are frequently used as distribution cables as they have the advantage of being flexible and so easier to install within the confines of buildings. They are also suitable for distribution circuits between buildings where the cables need to be installed in ducts or directly in the ground.

MIMS cable: may be used for distribution circuits where they supply safety services to ensure continued function in the event of fire.

Single-core cables in conduit and trunking may also be used for distribution circuits but the considerations of load and physical conductor size, particularly with respect to volt drop make these systems less common in all but the most compact of installations.

Try this

1   You are to carry out maintenance on the electical installation in a steel framed industrial unit which produces glazed ceramics for the hospitality industry. List the types of systems which could be used for each of the following:

a   a 16 kW kiln circuit

_____

b   socket outlets in the general office with solid floors

_____

c   rows of luminaires running the length of the unit.

_____

2 List the types of systems which could be used for each of the following:

a distribution circuits in a multi-storey office block

_____

b a supply to a separate building within a factory complex

_____

c the supply to a distribution board supplying firefighting equipment.

_____

Part 4 Management and alarm systems

Environmental control and building management systems

Environmental control and building management systems are used to control equipment within the building environment such as air conditioning, dust, odour and fume control and ventilation.

Many of these systems operate using twisted pair cables in a variety of forms. The type of cable and the csa of the cores will affect the distance from the control unit to the furthest point on the installation. For example a typical 0.2 mm^2 U/UTP cable would be suitable for a maximum length of 100m.

There are a number of different types of cable and each is classified by its construction. Table 8.1 identifies the cable type and the screening used for both the cable and the individual pairs.

Table 8.1 *Types of twisted pair cables*

Cable type	Cable screening	Pair shielding
U/UTP	none	none
U/FTP	none	foil
F/UTP	foil	none
S/FTP	braiding	foil
SF/UTP	foil, braiding	none

The abbreviations used in the cable types are:

- TP: twisted pair
- U: unshielded
- F: foil shielding
- S: braided shielding.

Most of the cables have a plastic insulation and sheath. The type of cable used depends upon the type of circuit and the level of segregation required.

The types include CY, SY and YY, braided flexible and local area network (LAN) cables all of which may be used in the environmental and building management control systems.

> **Note**
> There is more information regarding cables in the Planning and Selection study book in this series.

Emergency management systems

There are some basic circuits which apply to most types of alarm and call systems and it is worthwhile considering the basic circuits here.

Open circuit systems

The simplest circuit is that used on front door-bells, for it only contains one push, one sounder, a source of supply and the cable to connect them together. The push completes the circuit when it is pressed and is known as a 'push-to-make' type. The sounder may be a bell, chimes or a buzzer, but for simplicity we will continue to use the bell symbol.

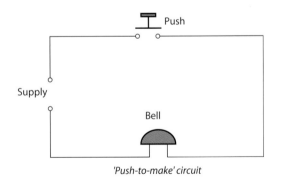

Figure 8.19 *Push-to-make circuit*

If an additional push is required in this circuit it must be connected in parallel with the first one.

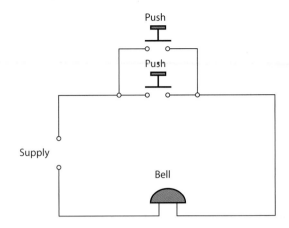

Figure 8.20 *Two call points in parallel*

In push-to-make circuits contact switches and other detection devices may be used to set off an alarm. Whilst this type of circuit has many uses its main disadvantage for alarm systems is that if a conductor is cut or broken such that the system is out of action.

Closed circuit systems

A closed circuit system involves the use of a sounder circuit and is the type of circuit primarily used for alarm systems. This means that the whole detection circuit is complete until something breaks it which sets off the alarm. To make the circuit practical a relay is used to divide the detection circuit from the alarm sounder circuit.

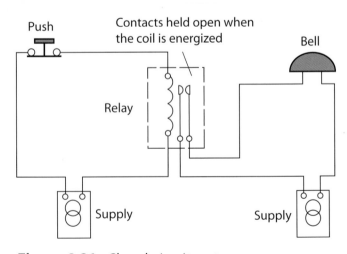

Figure 8.21 *Closed circuit system*

This relay is effectively the switch that closes the sounder circuit when the alarm is to be sounded. When the circuit is healthy the relay contacts are held apart by the magnetic field of the relay coil. This coil is energized all the time the detector circuit is complete. As soon as any part of the detector circuit is opened the alarm is sounded. Detectors, sensors and alarm switches in this circuit are connected in series.

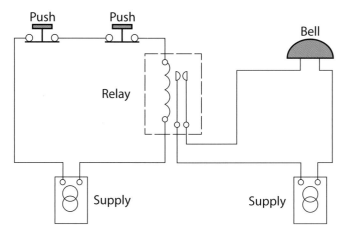

Figure 8.22 *Two detector points in series*

These normally closed circuits are often used for intruder and fire alarms. The detectors on contacts on doors, pressure pads, movement sensors or many other devices. On fire alarms they may be heat detectors, smoke detectors, flame detectors, rate of temperature rise sensors or straightforward manually operated break glass contacts.

End of line diodes and resistors are used with alarm circuits to allow the condition of the circuits to be monitored and to give an early warning of any faults which may occur on the system.

Figure 8.23 *Standard break glass unit*

Try this

Draw a circuit diagram for:

1   A normally closed circuit with two call points.

2   **A normally open circuit with three call points.**

Part 5 Emergency systems

Emergency lighting

Emergency systems include emergency lighting, which have a number of system options available:

- Emergency lights which have their own emergency power supply
- Emergency lights which have a central supply.

With both these options there is a choice of either:

- Maintained emergency lighting which is illuminated all the time.
- Non-maintained emergency lighting which is only illuminated when the main supply fails.

Central supply system

This may consist of a central battery bank with banks of secondary cells which are constantly on charge. In a very large installation the central supply could be a standby generator. Either method has its own distribution system and circuits wired through to where the emergency lights are required.

Local supply systems

It is often more practical to have special luminaires that have their own power source, i.e. self-contained luminaires. These can then be wired into the normal lighting supply circuit. These luminaires consist of a small battery-charging unit, batteries, relay and lamp. The batteries are constantly on charge all of the time the mains circuit is working correctly. When the mains supply fails the internal batteries take over. This system has the advantage of the emergency lights operating in the event of a local protective device operation without the need for additional control and monitoring circuits. However the regular testing of the emergency lights may take much longer as each light has to be tested individually.

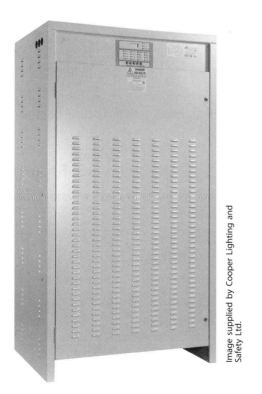

Image supplied by Cooper Lighting and Safety Ltd.

Figure 8.24 *Battery unit for centrally controlled emergency services*

TLC Direct

Figure 8.25 *Emergency luminaire with its own batteries and charger*

Maintained lighting

In public areas, such as theatres, the emergency signs have to be illuminated all of the time. Usually in these emergency signs the lamps are supplied by the battery which, under normal conditions, is being constantly charged. When the mains supply fails the batteries continue to keep the lamps illuminated.

Non-maintained lighting

In a non-maintained circuit the lights are only used when the mains supply fails. In luminaires that have the batteries contained within them, a relay switches the lamp on to the battery when the supply fails.

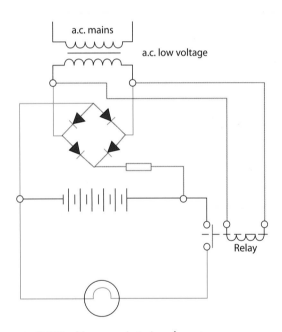

Figure 8.26 *Non-maintained system*

Fibre optics and light-emitting diode (LED) lighting is often used for emergency lighting systems. Fibre optics systems comprise fibre optic cables connected to a central projector containing a light source. The light source is connected to the emergency power supply. High intensity light from the lamps is transmitted down the optical fibres in the cable to the lighting points at floor or ceiling level.

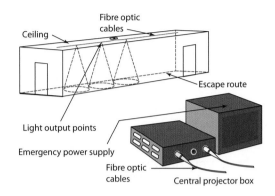

Figure 8.27 *Fibre optic emergency lighting system along a corridor*

The advantages of this type of system are:

- There is no current flow in the fibre optic cables and so no heat is produced and there are no magnetic fields generated around them. This makes them suitable in areas where:

    - Electromagnetic disturbances are to be avoided (data and signal transmission, etc.)
    - There is a risk of fire or they are situated in explosive atmospheres
    - They are in exposed or hostile environments without risk as the fibre optic cables are waterproof.

- A single light source is required for a large area and the type of lamps used provide long life for the system and makes maintenance easier as the light is in a single accessible location.

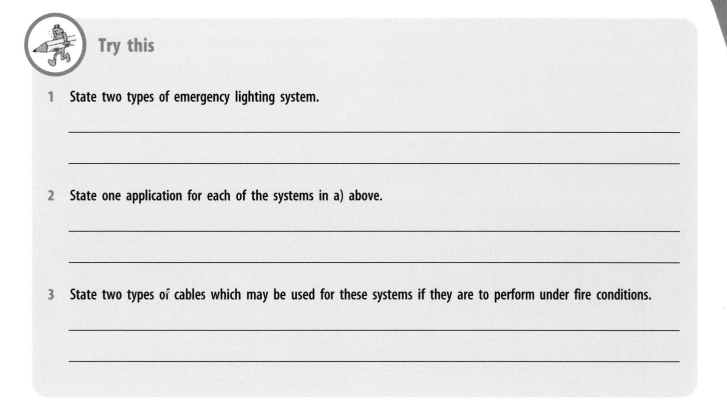

Try this

1 State two types of emergency lighting system.

_____

_____

2 State one application for each of the systems in a) above.

_____

_____

3 State two types of cables which may be used for these systems if they are to perform under fire conditions.

_____

_____

Part 6 CCTV, communication and data transmission systems

Closed Circuit TV, communication and data transmission systems

The types of cables used for these systems include Coaxial cables, Data cables- CY, SY and YY, braided flexible, LAN cables. The cable types used for data transmission systems are relevant to their particular applications.

Table 8.2 contains brief details of the most common data cables and their general applications.

Figure 8.28 *CAN-bus cable*

Controlled Area Network (CAN) is a system used in industries where complex controllers are required such as programmable control systems, building control systems and industrial control systems. The system does not address the equipment on the system but the messages themselves and so can deal with a large number of items of equipment. The CAN-bus system features high transfer at speeds between 1 Mbits/second at around 40 m and 5 Kbits/sec at 10 Kilometres.

Various forms of twin twisted pairs are used in the control industry and we shall look at some of the variations here. The type used relates to the level of shielding required to reduce electromagnetic interference for the particular application.

Table 8.2 *Common Cat cable information.*

Category	Type	Frequency bandwidth	Applications	Information
Cat 3	U/UTP	16MHz	10BASE-T and 100BASE-T4 Ethernet	Not suitable for speeds above 16Mb/s. so mainly used for telephone cables
Cat 5	U/UTP	100MHz	100BASE-TX and 1000BASE-T Ethernet	Common in most current LANs
Cat 5e	U/UTP	100MHz	100BASE-TX and 1000BASE-T Ethernet	Enhanced Cat 5. Same construction as Cat 5, but with better testing standards
Cat 7	S/FTP	600MHz	Telephone, CCTV and 1000BASE-TX in the same cable and 10GB Base-T Ethernet	Four pairs, S/FTP (shielded pairs, braid-screened cable).
Cat 7a	S/FTP	1000MHz	Telephone, CATV and 1000BASE-TX in the same cable and 10GB Base-T Ethernet	Four pairs, S/FTP (shielded pairs, braid-screened cable).

The shielding can be applied to the individual pairs or collections of pairs, the whole cable may be screened (shield around all the pairs combined) and a combination of these is also used. The shielding is generally earthed in order for it to be effective and an earth conductor is normally included in the cable. So let's look at the various types.

Unshielded twisted pair (UTP)

Unshielded twisted pair cables are used in many telephone and Ethernet networks. The common colour coding for the core pairs consists of one of the cores in solid colour the other core the same colour and white so the pairs can be readily identified. On large telephone systems 25 pair cables are common whilst general domestic systems utilize a smaller number such as the four pair shown in Figure 8.29.

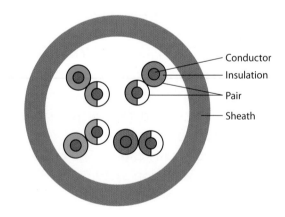

Figure 8.29 *Typical UTP cable.*

Telephone company distribution cables laid in the street may contain hundreds of pairs which are divided into smaller bundles each having a different twist rate which helps to reduce crosstalk.

Figure 8.30 *Typical UTP cable showing different twist rates.*

UTP cable is also used in computer networking data cables for short to medium runs because of its low cost compared with other options such as the fibre optic cables. It can also be used in security camera applications.

The other cable types which may be used and their construction are shown in Figures 8.31 to 8.33.

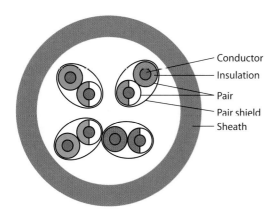

Figure 8.31 *FTP cable format*

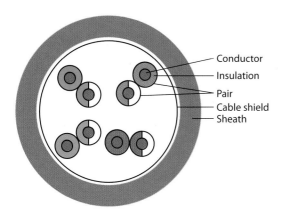

Figure 8.32 *F/UTP cable format*

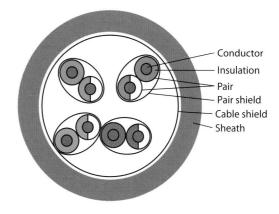

Figure 8.33 *S/FTP cable format*

Fibre Optic Cables

Many modern systems use fibre optic cables for these applications and their construction and advantages are as follows.

The conductor strands are made of optically pure glass as thin as a human hair (125 microns or 0.125 mm diameter), arranged in bundles (with many strands), and they may be used to carry digital information (in the form of light signals) over very long distances.

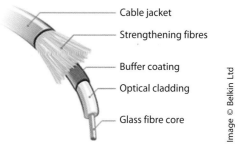

Figure 8.34 *Typical fibre optic cable construction*

Fibre optic cable is constructed with a core of thin glass fibre where the light travels, an outer cladding of optical material surrounding the core which reflects the light back into the core and an outer covering of plastic buffer coating which protects the fibres from damage and moisture.

The main advantages of fibre optic cable are that:

● It is less expensive than copper
● It has less signal loss than copper
● It has a lower power rating
● It is non-flammable – no electric current through fibres
● It is lightweight and flexible.

The type of cable to be used will depend upon the characteristics of the building, the requirements of the system and the desired performance. The specification for this will generally be provided by the system designer.

Equipment and accessories

When carrying out maintenance on electrical equipment consideration must be given to the environment in which the equipment is located and external influences to which it will be subjected during operation. Equipment and accessories used in adverse conditions such as wet and corrosive locations may deteriorate as a result of their everyday use. For example equipment installed in a car wash is subject to vibration, water and cleansing chemicals.

The information provided in the specification and the manufacturer's information needs to be consulted in the first instance to determine the requirements for the equipment and any replacement parts. During the maintenance activities we should determine whether the equipment is suitable for the location and duty being performed.

Excessive wear or the presence of excess dust or moisture in the equipment can indicate that the equipment is either:

● Not suitable for the environmental and/or operational conditions
● Failure of part of the equipment allowing these conditions to occur.

If the former is the case then a recommendation should be made to the client that a more suitable alternative is sourced and installed.

If the latter is the case then suitable maintenance and replacement of the defective part(s) should rectify the situation.

 Try this

1   **Explain the difference in the screening and shielding characteristics between a UTP cable and a S/STP cable.**

_____

_____

2   **State the common colour coding method for a typical UTP cable used for a telephone network.**

_____

_____

3 **State the purpose of the outer optical cladding in a fibre optic cable.**

_____

_____

Part 7 Work methods and procedures

When undertaking maintenance it is important that appropriate work methods and procedures are used. The requirements for the replacement and refitting of components within electrical systems and equipment are covered in Chapter 4 of this study book.

Many of the methods and procedures employed are going to be similar for any electrical maintenance activity so we will consider the common basic requirements.

The first consideration is to ensure that the procedures used ensure the safety of us and others whilst enabling the work to be carried out with the minimum of disruption to the client. The method of completing the work will, in some part, depend on the procedures that are followed and on a logical and efficient approach.

Systems

When working on any of the electrical systems including:

- Three-phase four-wire distribution systems
- Low voltage single and multiphase circuits
- ELV
- Lighting systems
- Heating and ventilating systems

the methods and procedures will be very similar.

Identify the extent of isolation that will be necessary to carry out the maintenance safely and agree with the client when this can be undertaken. Whilst some maintenance work will require the whole installation to be isolated other maintenance work can be safely and securely undertaken with local isolation. For example:

- Maintenance of the main switchgear for an industrial installation will require the isolation of the complete installation to allow this to be carried out safely. In some circumstances the DNO may need to be involved to remove and replace their supply fuses so that the main switchgear can be fully maintained
- Maintenance of a single item of machinery in the same location may be achieved by the safe isolation of the local isolator. However should the local isolator require maintenance then it will be necessary to isolate the circuit supplying the isolator.

When maintaining lighting systems it is important to determine the extent of the maintenance and in many cases the supply for the circuit will need to be safely isolated and secured in order to prevent danger.

It is important to remember that when undertaking maintenance of lighting systems there may be a need for temporary task lighting so that we can carry out the maintenance work safely. In addition, agreeing the isolation periods with the client may need to include the need for any temporary lighting to minimize the disruption to the client's business. Security may also be an issue if storage areas are left unlit during the work particularly where these may be accessed by others.

Often maintaining the lighting of one circuit at a time can minimize the disruption and keep the temporary lighting requirements to a minimum. Task lighting will always need to be considered. Working out of hours may reduce the disruption to the client's business but will often involve the need for more task and security lighting as a result.

Heating and ventilation systems present their own requirements during maintenance depending on the type of installation and the installation method. In some installations it may be possible to maintain individual heating and ventilation units. In others a centrally controlled and regulated system may require a full shutdown.

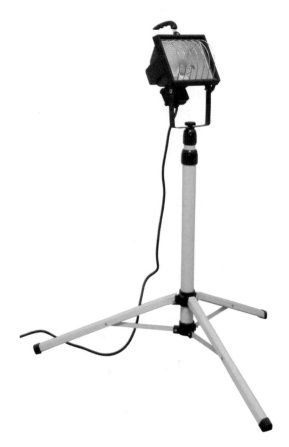

Figure 8.35 *Temporary lighting may be required*

Figure 8.36 *Maintaining a roof mounted chiller*

The environmental conditions will also have an impact on the method of working. For example consider a ventilation system installed for the care and welfare of livestock at say a chicken farm. The effects of the whole ventilation system being shut down can be both life threatening and very rapid. For this reason the system is often installed so that individual sections of the system can be maintained separately. Where the whole system has to be shut down sufficient staff and spare parts should be available to minimize the time taken and a temporary ventilation system may be necessary.

Figure 8.38 *Belt drive*

Maintaining drive systems will involve a similar approach to the maintenance of electrical equipment. BS 7671 identifies the requirements for switching for mechanical maintenance and this will require switching off and securing the drive. This is done by means of isolating the power source whether this is electrical, compressed air or hydraulic. The procedure for agreeing the timing and isolation of the system will be the same as that used for the electrical systems.

Figure 8.37 *Maintaining an office air conditioning unit*

Similar considerations apply to air conditioning and refrigeration systems. The comfort of personnel is one consideration but many of these systems are installed to maintain the condition of the client's stock. Chilled food storage facilities in a supermarket must be maintained within health and safety guidelines. If the temperatures go outside these guidelines the stock must be destroyed and the subsequent cost can be considerable. The need for the shutdown of these systems must be discussed with the client and wherever possible individual items of equipment maintained separately. Any major shutdown should be for the minimum time possible.

Figure 8.39 *Drive coupling*

The maintenance of security systems requires careful consideration as during this process there may be a need to isolate the system or part of it. Part of the maintenance for security systems and safety services requires checking the function of the system. However during the maintenance period any other maintenance such as the replacement of a sensor will mean that part or all of the system will be out of action.

Timing of the maintenance, particularly those which require battery duration tests, is critical as in the period immediately following the maintenance the batteries will still be in a discharged state. This means that should the system then be called into service the required duration may not be achieved.

Earthing and bonding systems requiring maintenance need to be carefully considered. Any maintenance which requires the disconnection of any of the conductors forming part of the earthing or bonding systems requires the isolation of the whole installation. The earthing and bonding together form part of the essential fault protection for the installation. Removal of any of these conductors without isolation of the supply presents a real risk of electric shock.

Data systems present particular problems when we are undertaking maintenance. The key requirement is to ensure that we have arranged with the client for the system, or the part of it we are working on, to be shut down and isolated. In addition we must confirm that the system has been shut down and it is ready for us to work on.

The maintenance needs to be carried out in a logical and efficient way to minimize the downtime. The testing and refurbishment of any data equipment must be carried out without damage to the equipment or installation.

The data system most people imagine is a computer system but it is important to remember that data systems operate over a wide range of facilities, equipment and systems. Data systems include the data controls for other services, machine controls and lighting. Some safety services use data systems as part of their operational and monitoring functions.

Equipment

The maintenance of equipment relates to the components and accessories used in the electrical and mechanical systems. These components include:

- Electrical plant such as transformers
- Motors and starters
- Switchgear and distribution panels
- Control components
- Contactors
- Power transmission mechanisms
- Luminaires and lamps
- Drive systems.

Figure 8.40 *Checking the brake setting on lifting gear*

Essentially we will follow the same procedures for the maintenance of equipment as we did for systems. It is important to make sure that the necessary equipment, replacements and spare parts are available before work begins. It is also important to coordinate the maintenance of the equipment with the relevant parts of the system. If the system is to be isolated then maintaining the equipment at the same time will reduce the inconvenience for the client. In some instances it will not be possible to maintain all of the equipment at the same time. Where this is the case the sequence of maintenance for the equipment should be such that the disruption is kept to the minimum.

We need to refer to the manufacturer's information, the specification and diagrams and drawings in order to ensure the correct maintenance is carried out using appropriate spare parts.

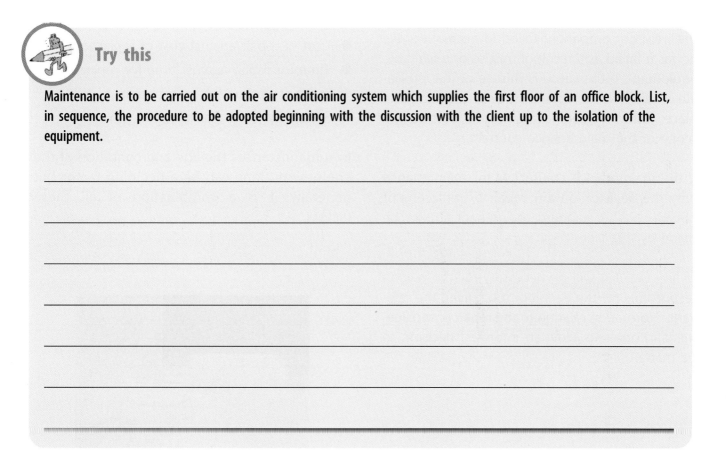

Try this

Maintenance is to be carried out on the air conditioning system which supplies the first floor of an office block. List, in sequence, the procedure to be adopted beginning with the discussion with the client up to the isolation of the equipment.

_____

_____

_____

_____

_____

_____

Part 8 Corrective actions and repairs

We have looked at the requirements for maintenance and the need for the normal replacement and spare parts required during this process. It is not unusual during maintenance activities for additional actions or repairs, which are not part of the normal maintenance activities, to be identified. We need to consider the appropriate actions where these situations occur, beginning with who has the responsibility for making the decisions.

The responsibility for these decisions will depend upon the contractual arrangements and ownership of the equipment or system.

If your company's contractual arrangement for the maintenance includes both the maintenance and replacement of defective parts then the decision will be with your company and in the first instance the information will be passed to your supervisor for a decision. Your supervisor will need to determine whether the necessary work is as a result of defective parts and will require your input to make this decision. Essentially, has the item or component failed in normal service or has it failed as a result of misuse or unsuitable equipment. If the cause is misuse or inappropriate equipment a discussion will need to take place with the client as it may be outside the scope of the maintenance contract.

If the maintenance contract is for maintenance only the decision on any repair or replacement outside of the standard maintenance must be made by the client.

In any event it will be necessary for the person responsible for the decision to be given the relevant information relating to the repair or replacement in order to allow an informed decision to be made.

Cause and effect

The considerations here are:

- What was the cause of the failure and will the simple component replacement cure the underlying cause
- Will the sourcing and replacement of the item involve more time and cost in labour and lost production
- Would the replacement, of say a printed circuit board, be cost-effective and reliable.

Viability

The viability of repairing faulty equipment should be discussed once the availability and relative costs have been established. The basic formula for determining the viability of repair or replacement includes:

- Overall age and condition of equipment
- Cost of repair against cost of replacement
- Time for repair against time for replacement
- Total downtime for each option
- Suitability for the task.

In some instances the age and condition of the equipment alone can be a deciding factor but generally it is a combination of all these factors.

Figure 8.41 *Condition can be a deciding factor for repairs*

Replacement parts

Are the necessary replacement parts readily available and if not what is the likely period for them to be obtained.

For some faults there is little difficulty in obtaining replacement parts, such as general purpose cables, accessories and the like. There may some availability issues however with some of the more specialist items such as replacement individual components or a printed circuit board.

Downtime

One of the key considerations when repairing faults is the period for which the equipment, circuit or installation will be out of service. The effects of inconvenience and lost production can be considerable and there are a number of important factors which need to be determined.

- Can the repair or a replacement significantly reduce the downtime?
- Can the item of equipment or system be out of commission for the period without affecting the activity of the client?
- Which of a repair or a replacement can be carried out quickest?

Cost

The main factors which affect the cost of repairs are:

- Spares and materials
- Downtime
- Accessibility

- Equipment
- Labour.

Note

There is more information regarding the implications of repair or replacing in the Fault Finding and Diagnosis study book in this series.

Implications for the repairer

Once an agreement to carry out the repairs has been made there are some implications for those carrying out effective repairs. The client may well expect to have a program of work and the effect on their business operation will be measured against this. Should the repair not be completed in time the client may well seek compensation for any additional costs or loss of revenue involved.

It would be reasonable for the client to expect some form of warranty for the repair from the contractor and this needs to be considered and discussed before the work is undertaken. For example a repair to an item of equipment may carry no manufacturer's guarantee or warranty. The replacement of a complete item would attract the standard manufacturer's warranty and so minimize additional cost to the contractor.

The period for which the repair is guaranteed and the extent of the guarantee is significant particularly where there is a risk of claims for loss of production following a further failure.

Carrying out repairs

It may be possible to carry out these additional repairs at the same time as the scheduled maintenance. This is generally providing that:

● The agreement for the work is given promptly
● The spare or replacement parts are available
● Suitable manpower is available.

The maintenance program may need to be completed to allow the rest of the system or equipment to be placed back in service before the repair is carried out. This will help to minimize the disruption to the client.

Where this is not the case the safety and implications of leaving the item of equipment or system in service must be considered.

If there is immediate danger from leaving the item in service the client must be advised that the item remains out of service until the repair is effected. Timing of the repair must be agreed with the client if this is to be carried out at a different time to the maintenance.

The repair procedure will be similar to that used for the maintenance activities only this will relate to the specific item(s) to be repaired or replaced.

Try this?

During the routine maintenance of an electric motor on an assembly line it is found that the flexible metal conduit between the starter and the motor has been damaged midway between the two and the conductors are exposed. The flexible conduit is not repairable.

Produce a list of the items that need to be considered and made available to the client to enable them to determine the action.

_____

_____

_____

_____

_____

Task

List the actions that would be required to safely carry out the replacement of the metal flexible conduit.

_____

_____

_____

_____

_____

_____

_____

_____

_____

_____

_____

_____

_____

_____

_____

_____

_____

_____

_____

_____

_____

_____

_____

Try this: Crossword

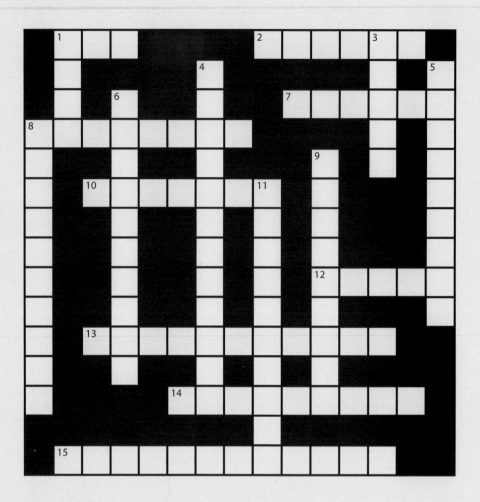

Across

1   Tin ... (but they're often aluminium) (3)

2   Switching from two locations is this type of circuit (3–3)

7   Following an equipment failure this may be required (6)

8   System used to detect intruders (8)

10   This may be circuit or wiring (7)

12   Volts x Amps is the apparent version of this (5)

13   If we cannot do 7 across we may need to do this (11)

14   A component which is not functioning correctly is this (9)

15   Who is responsible for making the equipment? (12)

Down

1   One of these is required for the system in 8 across (4)

3   Without 1 down we may sound this (5)

4   Comes between the switches in 2 across (12)

5   The way in which a process is carried out (9)

6   Used in most installations to provide light (10)

8   This is installed to control installations or equipment (10)

9   This refers to items connected to the installation (9)

11   The duty holder has a responsibility for the ... of the maintenance for the installation (10)

Congratulations you have now completed Chapter 8 of this workbook. Complete the self assessment exercise before you progress to the next chapter.

SELF ASSESSMENT

1 The information required for the routine maintenance of an item of equipment may be found in:

  a. BS 7671

  b. The Electricity at Work Regulations

  c. The manufacturer's instructions

  d. The Health and Safety at Work (etc.) Act

2 Where a light is to be controlled from both ends of a corridor and at the midpoint the type of switching to be used is:

  a. Three one-way

  b. Three two-way

  c. Two two-way and a one-way

  d. Two-way and intermediate

3 The purpose of the coil in Figure 1 is to:

  a. Draw the contacts together

  b. Make the bell ring

  c. Isolate the bell circuit

  d. Hold the contacts open

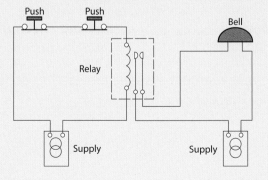

Figure 1

4 The type of cable shown in Figure 2 is:

  a. S/FTT

  b. F/UTP

  c. UTP

  d. FTP

Figure 2

5 One advantage of fibre optic cables over copper conductors is that they:

  a. Are more rigid

  b. Have a lower power rating

  c. Have a higher power rating

  d. Are more expensive

Procedures and documentary systems

RECAP

Before you start work on this chapter, complete the exercise below to ensure that you remember what you learned earlier.

The manufacturer's information contains information for maintenance including _____ and circuit diagrams, _____ and adjustments for _____ operation and _____ finding charts.

Flat multicore and _____ cables are suitable in locations where they are not _____ to _____ conditions or _____ damage.

Conduit and _____ containment systems providing _____ protection for _____ core cables allows the use of _____ routes for a number of _____ and the _____ of existing circuits.

Modular _____ systems provide a _____ system with the facility to _____ sophisticated _____ systems.

Powertrack and _____ systems are a prewired _____ trunking which uses copper _____ conductors and _____ units to tap off _____ where it is needed.

Environmental control and _____ management systems are used to control air-_____, dust and _____ extraction and _____.

_____ -circuit wiring systems are _____ used for alarm circuits.

Emergency lighting may be installed as either a _____ or _____ system.

Controlled _____ Network is used where _____ controls are required such as _____ control systems and_____ control systems. This system addresses the_____ themselves and so can _____ with a large _____ of items of _____ .

Unshielded _____ -pair cables are used in _____ systems and the common colour coding for core pairs is _____ of the cores in _____ colour the other core the same _____ and _____ .

Fibre optic cable is constructed with a _____ of thin _____ fibre, an _____ cladding of _____ material and an outer_____ of plastic buffer coating.

Excessive _____ in equipment can indicate that the equipment is not _____ for the _____ or _____ conditions.

When undertaking maintenance it is important to ensure that the _____ used ensure the _____ of us and others minimizing _____ to the client.

When maintaining data systems _____ must be made with the _____ for the system to be _____ and isolated.

It is important for the client to be given the _____ information relating to the repair or _____ in order to allow an informed _____ to be made.

Factors which affect the cost of repairs include downtime, _____ , equipment and _____ .

LEARNING OBJECTIVES

On completion of this chapter you should be able to:

● State the reasons why it is necessary to undertake regular inspection, adjustment and replacement of different electrical systems and equipment

● Specify the documentary processes and procedures that are necessary for electrical maintenance work, including:

– Workplace requirements for, and the importance of:

– Documenting information

– Reporting findings and variations from the maintenance schedule.

– Procedures for the completion of necessary maintenance documentation – including appropriate organizational or external Quality Assurance systems

– Interpret technical and functional information and data from maintenance logs.

Part 1 Reasons for maintenance

This chapter considers the procedures and documentary systems related to maintenance activities. Whilst working through this chapter you may wish to refer to BS 7671, Requirements for Electrical Installations, IET Guidance Note 3, Inspection and Testing and the maintenance schedules, logs and reports used by your company.

Remember

Regulation 29 of the Electricity at Work Regulations requires us to be able to demonstrate that all reasonable care has been taken to prevent danger arising. The maintenance records are one method of demonstrating this.

The need for maintenance

Regular inspection, adjustment and replacement of different electrical systems and equipment is necessary for a number of reasons including:

- Legislative requirements
- The prevention of danger
- To reduce costs
- To maintain reliability
- Environmental requirements – to prevent pollution.

Legislative requirements

As we have seen earlier in this study book there are a number of statutory requirements for the equipment used in the workplace and public buildings to be maintained so that it is safe for use. Whilst this legislation does not specify regular inspection for the equipment this, together with records of the inspection, testing and any repairs, is the most logical method of demonstrating that the statutory requirements for maintenance have been met.

The prevention of danger

During maintenance, systems and equipment are monitored and, when necessary, repaired in order to keep them in a safe condition. We are required by law to do this to prevent danger to persons, livestock and the environment. Without maintenance, equipment and systems cannot be kept in a safe condition. Maintaining systems and equipment involves:

- Inspection and/or testing
- Maintaining records of the results
- Monitoring the maintenance procedure.

During the inspection and testing, repairs may be identified which are necessary to keep the system and equipment in a safe condition.

Reducing costs

When equipment fails or malfunctions this can result in loss of service, loss of production and the need for repairs, all of which can involve a considerable cost to the client.

The cost of the repair is dependent on a number of factors including:

- Are the necessary replacement parts readily available? If they are not then the cost due to loss of service and production will rise
- Is repair or is replacement the most cost effective option? The lack of spares, the age of the equipment and the time involved will all have an effect on the overall cost
- When can the work be carried out to minimize further disruption and inconvenience to the client? If the complete installation or part thereof has to be isolated the disruption to production will be increased
- Are sufficient skilled staff available to carry out the necessary repair work? The availability of skilled staff will determine the period for which the equipment or system will be out of action. The longer the wait the more production will be affected.

Regular planned preventative maintenance can significantly reduce the likelihood of failure and so reduce the cost and disruption for the client. This in turn helps to maintain the reliability of the system.

Maintaining reliability

It is important that systems and equipment are maintained to keep them operating safely and effectively. There are some systems which are essential such as fire alarm, security systems and emergency lighting. However, reliability is important for manufacturing and business processes and failures will usually result in loss of revenue. In manufacturing, the cost may be high as loss of production and wages to employees add to the potential loss of orders. In the data industries reliability is paramount and any failure can have a catastrophic effect on the business.

Therefore regular maintenance will minimize the chances of a failure and improve the reliability of systems and equipment.

Figure 9.1 *Maintenance for better reliability*

Environmental requirements

There are a number of systems and processes which are essential to meet the environmental legislation and other requirements. Regular maintenance is important to ensure that these systems and the associated equipment continue to provide the necessary environmental protection.

Legislation such as the Environmental Protection Act and the Clean Air Act are in place to protect the environment. Failure to comply may result in severe penalties and may result in cessation of the work activities involved. Typically, processing and filtration systems for air and water quality and other products which are discharged or disposed into the environment will need to be fully maintained.

Note

Under the Environmental Protection Act pollution is defined as

'A release of any substance into air, water or land as a result of any process which causes harm to man.'

Try this

List five reasons why maintenance should be carried out.

1 _____

2 _____

3 _____

4 _____

5 _____

Part 2   Documentation

Documentary processes and procedures

As we discussed at the beginning of this chapter there are a number of statutory regulations which require systems and equipment to be maintained in a safe condition to prevent danger. Whilst none of these detail the procedures or require the keeping of records, without documentary evidence it is extremely difficult to demonstrate compliance with these regulations.

The keeping of maintenance records is advisable to:

● Prove that maintenance has taken place
● Show the condition of the system and equipment
● Allow systems and equipment to be monitored.

Maintenance work on systems and equipment includes the requirement for records to be kept throughout their working life. It is the responsibility of the person responsible for the system and equipment, the Duty Holder, to keep and maintain these records.

Recommended records to be kept for the in-service inspection and testing of electrical equipment include:

● A register of all equipment
● Formal inspection records
● Combined inspection and test records
● Faulty equipment records
● Repairs record.

Where fixed equipment is subject to maintenance then a maintenance record and schedule should be maintained detailing the frequency of the maintenance, what maintenance is required on each occasion and the results of the previous maintenance.

Where periodic inspection and testing is carried out on electrical installations (this forms part of the maintenance for the installation) and fixed wiring are carried out the results are recorded on:

- An Electrical Installation Condition Report together with
- Schedule(s) of Inspections
- Schedule(s) of Test Results.

The Duty Holder should make the results of the previous maintenance, inspections and testing available to the person carrying out the maintenance. Without this information the current condition and test results cannot be compared to the earlier condition to determine whether deterioration has taken place.

When carrying out periodic inspection and testing of electrical installations it is acceptable to carry out sample inspection and testing in order to minimize the interference with the installation. If the sampled areas show no sign of deterioration and are typical of the installation then further sampling may not be necessary. However, where the previous records are not available this sampling procedure would not be suitable and a full inspection and test may be required.

It is important that the maintenance records are accurately recorded as this allows the Duty Holder to make informed decisions on the action that needs to be taken. Maintenance schedules will normally include sections for recording the findings, condition and test result information. Our responsibility when carrying out maintenance is to make sure that:

- The correct procedures are carried out
- The results of tests are accurately recorded
- The condition and any additional findings are recorded.

Any variations or changes from the maintenance schedule must be recorded together with the reasons for the variation. There are occasions where changes are made due to new technology or materials. For example: the wet cell batteries of a small centrally controlled emergency lighting system have been replaced by sealed-for-life gel-filled batteries. This will result in a change to the maintenance and the checks for electrolyte specific gravity and levels will no longer be necessary.

Similarly, replacement components may also have a different maintenance requirement to the original and so the maintenance schedule will alter. The manufacturer's information for these replacement components or items of equipment should be passed to the client and kept with the maintenance information.

There are, as we discussed in the previous chapter of this study book, occasions when other conditions are established during the maintenance activity and it is important that this is relayed to the Duty Holder. For example whilst maintaining a roof mounted air conditioning unit corrosion to the enclosure and metallic conduit is observed. This may not affect the maintenance or operation of the unit but left untreated will get worse and eventually breach the enclosure and expose the cables within the conduit. This would allow water ingress to these areas and could result in failure of the unit. Notifying the Duty Holder of the condition will allow action to be taken to clean and protect the affected areas and so extend their serviceable life.

Whenever and wherever maintenance is carried out we should report findings related to the maintenance and any other conditions we observe whilst carrying out the work. We are not expected to go looking for problems but any we do see should be reported to the Duty Holder.

Figure 9.2 *Corrosion needs to be reported*

Some of the areas which we may consider during maintenance activities include:

- Corrosion of enclosures, fixings, etc.
- Damage to apparatus or wiring systems
- Excessive accumulation of dust and dirt
- Loose electrical connections
- Loose fixings, glands, conduit, stoppers, etc.
- Conditions of enclosure gaskets and fastenings
- Oil or compound leaks
- Bearing conditions
- Inadvertent contact between rotating and fixed parts
- Integrity of guards
- Ratings and types of lamps
- Vibration
- Correct working of relays and protective devices
- Unauthorized modifications or adjustments
- Maintenance carried out other than to manufacturers' recommendations.

Any of these which are not rectified as part of the maintenance should be reported. Some items require little effort, time or spare parts and so can be readily fixed during the maintenance. However these should still be recorded as there is likely to be an underlying cause for the condition which should be investigated. If the situation, although rectified, is not reported and recorded no preventative action will be undertaken.

Periodic inspection of electrical installations

The report produced following a periodic inspection records the condition and compliance of an existing electrical installation, which has been in use, with the current requirements of BS 7671.

Remember

A periodic inspection considers the compliance of an existing installation with the current requirements of BS 7671, irrespective of the age of the installation or what regulations were effective at the time the installation was completed.

Every electrical installation deteriorates with age and use. It is important to ensure that the installation continues to be safe for use and does not put users of the installation at risk. The frequency of the inspection is dependent upon a number of factors, such as the type of building and its use, the age of the installation, and the environmental conditions. IET Guidance Note 3 provides some guidance on the maximum period to the first inspection and test of a new installation based upon these criteria. The frequencies between inspections are determined considering these factors, and the condition of the installation following a period of use.

The purpose of the periodic inspection is to establish, as far as is reasonably practicable, any factors which could impair the safety of the electrical installation and report on them. Chapter 62 of BS 7671 outlines the requirements for periodic inspection and testing of electrical installations. Model forms for recording the findings of a periodic inspection and test are given in Appendix 6 of BS 7671 and Part 5 of IET Guidance Note 3.

Where a planned preventative maintenance scheme which includes the fixed wiring is in operation and the inspection and test results are recorded there may be no requirement for a periodic inspection and test to be undertaken.

Where the fixed wiring is part of the maintenance scheme the test results need to be recorded and the standard form for the Schedule of Test Results provided in BS 7671 may be used. This form contains the minimum information that is required in terms of the recording of the results. Similar forms may be used but should contain at least the information required on the standard form. Some companies use schedules which contain additional information to the standard form and this is acceptable.

Figure 9.3 shows a typical Schedule of Test Results and the extent of the testing and the number of schedules required depends on the complexity of the installation and the number of circuits included in the maintenance process.

It may be that only a small number of circuits are involved at each maintenance stage and a schedule of test results for those circuits is completed and retained with the maintenance records.

Note

Further information on the testing and completion of the schedules is contained in the Inspection and Testing study book in this series.

SCHEDULE OF TEST RESULTS Sheet [6] of [6]   ⊛ECA

DB Reference no. DB F1
Location First floor lobby cupboard
Zs at DB (Ω) 0.6
I_pf at DB (kA) 0.48
Correct polarity of supply confirmed YES / ~~NO~~
Phase sequence confirmed (where appropriate) N/A

Details of circuits and/or installed equipment vulnerable to damage when testing

Details of test instruments used (state serial and/or asset numbers)
Continuity MF 1006 JD
Insulation resistance **
Earth fault loop impedance **
RCD **
Earth electrode resistance N/A

Tested by:
Name (CAPITALS) JAMES DOUGLAS
Signature James Douglas   Date – July 20–

Circuit number	Circuit description	BS (EN)	Type	Rating (A)	Breaking capacity (kA)	Reference method	Live (mm²)	cpc (mm²)	r₁ (line)	r_n (neutral)	r₂ (cpc)	(R₁+R₂)*	R₂	Live–Live	Live–E	Polarity ✓	Ω	@I_n	@5I_n	Test button operation	Remarks (continue on a separate sheet if necessary)
A	B	C	D	E	F	G	H	I	J	K	L	M	N	O	P	Q	R	S	T	U	V
1	Print Room Power	61009	C	40	6	100	10	4.0	N/A	N/A	N/A	0.16		199	99	✓	0.75	45.5	6.8	✓	
2	Staff Kitchen Power	61009	B	32	6	100	10	4.0	N/A	N/A	N/A	0.03		100	100	✓	0.63	38.6	11.2	✓	
3	Front Ring Circuit	61009	B	32	6	100	2.5	1.5	0.45	0.48	0.95	0.3		30	28	✓	0.62	36.4	9.6	✓	
4	Back Ring Circuit	61009	B	32	6	100	2.5	1.5	0.60	0.62	1.03	0.41		50	48	✓	1.00	38.5	10.5	✓	
5	Boiler	61009	B	16	6	100	2.5	1.5	N/A	N/A	N/A	0.15		20	20	✓	0.75	40.0	11.8	✓	
6	Immersion Heater	61009	B	16	6	100	2.5	1.5	N/A	N/A	N/A	0.46		52	52	✓	1.06	42.0	15.3	✓	
7	Kitchen Freezer socket	60898	B	16	6	100 & A	2.5	1.5	N/A	N/A	N/A	0.19		200	200	✓	0.79	N/A	N/A	✓	
8	Front Lights	61009	B	6	6	100	1.5	1.0	N/A	N/A	N/A	1.5		36	50	✓	2.1	200	12.5	✓	
9	Rear Lights	61009	B	6	6	100	1.5	1.0	N/A	N/A	N/A	1.2		>999	955	✓	1.8	120	12.5	✓	
10	Spare																				

* Where there are no spurs connected to a ring final circuit this value is also the (R₁+R₂) of the circuit.

C-STR-ECA REV Aug 2011 V1

Figure 9.3 *Schedule of Test Results*

Completing documentation

Our responsibility is to complete the maintenance documentation accurately. The documentation required can vary from organization to organization and is normally either:

- Provided by the client using their own documentation
- Using documentation produced by the maintenance (our) company.

We have considered various styles of maintenance record forms earlier in this study book together with the information to be recorded but a recap here will be beneficial.

Figure 9.4 shows a typical maintenance record for maintenance on a lathe with the type of information that may be required. The information required will vary depending on the equipment involved and there may be several pages of information for major services on complex equipment.

Figure 9.5 is a typical maintenance record sheet for the annual maintenance on an air conditioning unit. This type of record is often used by companies as it:

- Details what work is required
- Acts as an aide-memoire for the maintenance engineer

- Minimizes the entry requirements from the engineer
- Standardizes the maintenance procedure and record.

As we can see there are more items to check and details to record and as the complexity and maintenance requirements increase so too will the information to be recorded. The maintenance records are to be passed to the client as they are responsible for keeping these records and ensuring they are up-to-date. These records form the maintenance log for the company and contain the information which is required at any maintenance process so that the ongoing condition can be monitored. As we already discussed the log will also contain the information relevant for the maintenance procedures.

Many maintenance companies and other organizations have their own quality assurance (QA) systems and these are often verified by an independent third party. The purpose of a quality assurance system is to ensure that a company is providing the best possible service or product. This includes a requirement that the same procedures are followed for an activity on every occasion. It does not necessarily mean that the product is of good quality but the same procedure is always followed and the end product is consistent as a result. If the process results in an unsatisfactory outcome then the process has to be changed.

Douglas Turning Ltd				Maintenance Record								
Brown and Becker Lathe Ref. N° 0194												
Date	Bearings		Starter contacts		OL Settings		Gear Box		Test Results				Function
	Condition	Greased	Clean	Replace	✓	Adj	Oil	Func	IR MΩ	Cont	Ø Seq	EFLI Ω	✓/✗
---/07/20--	Good no play	HP40	✓	No	✓	No	OK	✓	>299	✓	✓	0.9	✓
Name: James Douglas			Signature: *James Douglas*					Date: --/07/20--				

Figure 9.4 *Typical maintenance record*

Air Conditioning Annual Maintenance			
Date: .../..../....	Engineer:	Unit N°..............	
Task	Action	Result	Units
Fit refrigerant gauges			
Check compressor belt	Replace belt	Yes/No	
Check compressor	Mounting kit OK	Yes/No	
	Tensioner adjustment	Yes/No	
	Clutch bearing OK	Yes/No	
	Mounting bolts torque OK	Yes/No	
Check Hoses	Fittings OK	Yes/No	
	Wear/chafing evident	Yes/No	
	Secured OK	Yes/No	
Check Condenser	Clear of debris	Yes/No	
	Fan(s) Operating	Yes/No	
	Secured OK	Yes/No	
Check Evaporator	Blower operating	Yes/No	
	Clear of debris	Yes/No	
	Secure	Yes/No	
Pressure Switch	Low	....	bar
	Medium	....	bar
	High	....	bar
Electrical	Fuses OK	Yes/No	
	Connections:		
	Earth OK	Yes/No	
	Battery OK	Yes/No	
	Ignition OK	Yes/No	
	EFLI	....	☒
Controller Settings	Operates at :		
	High	Yes/No	
	Low	Yes/No	
	Fan speeds OK	Yes/No	
	Manual stat operation OK	Yes/No	
	Re-circ controls operation OK	Yes/No	
Leak check	Dye added	Yes/No	
Reclaim refrigerant		Yes	
Filter dryer replaced		Yes	
Pressure test system	20 minutes	....	bar
Evacuate system		Yes/No	
Recharge system	Refrigerant type	....	
	Charge	....	kg
	Oil added	Yes/No	
	Oil type	....	
Run and Test	Down to set point	Yes/No	
Remove refrigerant gauges			
Additional requirements			

Figure 9.5 *Typical air conditioning unit annual schedule*

For service companies this involves the company using the same procedures for their work activities and providing their client with suitable and correct information and records. The use of record sheets such as that in Figure 9.5 forms part of this quality process. Every air handling maintenance will have the same checks and the information provided to each client will be in the same format. There may be additional information that needs to be provided and records to be maintained by the company to satisfy their own QA procedure.

For example a company has a procedures' manual which details the activities and procedures to be adopted by the operatives. The company's QA procedure will often require that any changes, additions or updates to these procedures are issued to all the relevant staff. The manual may be in either hard copy or held electronically but all relevant staff must be able to access the manual. The QA procedure will require that there is a unique record of the issue of each change or addition to the staff and an acknowledgement from each operative that this has been received and acted upon. So the maintenance schedule may have some change made to it and the company QA process confirms that:

● Staff have been notified of the change
● Individuals have confirmed the receipt of the change
● The new schedule is the only one being used.

We may be involved in the completion of additional documentation in order for our employer's QA system to be administered and for them to continue to be a Quality Assured company.

Interpreting information

Before and after carrying out maintenance we need to consider the information contained in the maintenance logs for the system and/or equipment. The technical information and functional details will include such data as:

● Settings for operational limits including settings for pressure, temperature, travel start/stop and fan speeds
● Torque settings for fixings, bolts and terminations
● Tolerances allowed for bearings, sleeves and pistons
● Performance characteristics such as fan extract rates.

The maintenance log may also include information on measurement data such as voltage and current which are used during maintenance to confirm that equipment is performing correctly. This information is important as it allows the existing characteristics to be checked and where necessary adjusted during maintenance.

In addition to the technical and functional information available in the maintenance log there will also be reference to the frequency and results of previous maintenance activities. These give an indication of the results that we can expect from our own maintenance and provide a reference for comparison. We can use this comparison to establish trends and changes in condition.

Where results vary considerably there will be an underlying cause and this will need to be established. In some instances it is simply the aging of components or materials which affect the results. Every electrical installation and the components within it will deteriorate over time; it is an inevitable aging process. Some materials deteriorate quicker than others and some components are more susceptible than others. Other factors which determine the degradation of equipment and materials include their use and the environmental conditions.

The type of installation, the environmental conditions and the frequency of use will affect the equipment and system. As the process is different for individual locations the previous records are a useful means of determining the condition and effectiveness of maintenance.

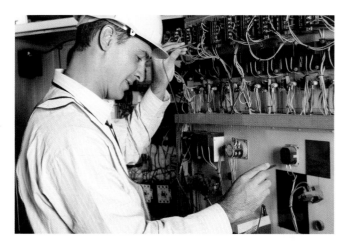

Figure 9.6 *Checking output data*

For example: maintenance is being carried out on a dust extract system for a powder coating booth. After cleaning the fan and housing, testing the fan speed controls reveals that the performance is below the acceptable extract values on the low and medium settings. Checking the previous maintenance data can indicate whether there has been a recent drop in performance. If the extract rates are shown as declining but within limits over say the previous two tests then there is a good indication that components may be failing. Further checks, using the technical data can be used to determine whether the output from the controller is correct (voltage etc.). If the results show the controller output to be below acceptable values the controller is likely to be the reason for the poor performance

and the client is advised that a replacement should be fitted.

Similarly the gradual fall in insulation resistance of a cable over time indicates the insulation is failing whereas a sudden fall may indicate a problem with the circuit.

In summary the use of the information and data in the maintenance log allows the engineer to confirm performance, check performance parameters and identify both current and potential problems. This information can then be passed to the client to make an informed decision on the action they wish to take.

The process of interpreting information and data also requires some experience of the particular system or equipment involved. The inspection and testing of fixed wiring is a particular example of this requirement.

Knowledge and experience of the type of installation, supply system and the process of inspection and testing is necessary to be able to carry out the inspection and testing of in-service electrical installations. This together with the requirements of BS 7671 is essential to enable the compliance and safety of the installation to be determined.

Try this

1   List three reasons why it is advisable to keep maintenance records.

a _____

b _____

c _____

2 Maintenance is to be carried out on a machine motor isolator/starter supplied via a steel wire armoured cable. List three additional areas which should be considered whilst carrying out the maintenance.

a _____

b _____

c _____

3 State the conditions that must exist in order that a periodic inspection and test of the fixed installation need not be carried out.

_____

_____

_____

4 List four items of information which should be contained in the maintenance log which can be used during the maintenance of an air handling unit.

a _____

b _____

c _____

d _____

Congratulations you have now completed the final chapter of this unit and study book. Complete the self assessment questions before you carry on to the end test.

SELF ASSESSMENT

1   Which of the following is **NOT** a reason for carrying out maintenance?

a.   Legislative requirement

b.   Increase production

c.   Maintain reliability

d.   Prevent danger

2   a) Maintenance is never required for environmental reasons

b) Regular maintenance can reduce costs

Which of the following is correct in regards to the two statements above?

a.   Statement a) is correct and statement b) is incorrect

b.   Statement a) is incorrect and statement b) is correct

c.   Both statements a) and b) are correct

d.   Both statements a) and b) are incorrect

3   Which of the following is one reason why maintenance records are kept?

a.   It is a legislative requirement

b.   It guarantees a reduction in insurance

c.   It demonstrates compliance with legislation

d.   BS 7671 requires all maintenance to be recorded

4   An electrical installation is subject to a planned maintenance regime. Which of the following is a further condition that would result in a periodic inspection not being required?

a.   Records of test results are kept

b.   Maintenance is carried out each year

c.   A separate service agreement is in place

d.   No machinery is installed in the building

5   The purpose of a quality assurance system is to ensure that a company is:

a.   Providing best quality products

b.   Not going to default on payment

c.   Registered with a third party regulator

d.   Providing the best possible service or product

End test

1. Which of the following is used to confirm safe isolation of a circuit?

   ☐ a. An ammeter

   ☐ b. A low resistance ohmmeter

   ☐ c. An approved voltage indicator

   ☐ d. A voltstick

2. Which of the following is the total number of tests to be made when confirming safe isolation of a single-phase circuit?

   ☐ a. 10

   ☐ b. 7

   ☐ c. 4

   ☐ d. 3

3. When maintenance is to be carried out and the equipment is energized, the first action to be taken is:

   ☐ a. Safe isolation

   ☐ b. To lock off the supply

   ☐ c. To obtain permission to isolate

   ☐ d. To isolate the supply and notify the client

4. One implication of carrying out safe isolation of a circuit is:

   ☐ a. The circuit will function normally

   ☐ b. It will take longer to complete the work

   ☐ c. There may be some inconvenience to the user

   ☐ d. There will be no effect on the use of the installation

5. Which of the following is the nominal supply voltage for U and U_0 on a three-phase system on the public distribution network in the UK?

   ☐ a. 415/240 V

   ☐ b. 415/230 V

   ☐ c. 400/140 V

   ☐ d. 400/230 V

6. In a balanced three-phase circuit the current in the neutral conductor will be:

   ☐ a. No current at all

   ☐ b. The load current

   ☐ c. Half the load current

   ☐ d. The resultant current

7. The line current of a three-phase load is 53 A. The current in the phase winding is:

   ☐ a. 17.6 A

   ☐ b. 31 A

   ☐ c. 92 A

   ☐ d. 159 A

8. During maintenance on a three-phase four-wire control system the voltage measured between the line conductors is 440 V. The measured line to neutral voltage will be:

   ☐ a. 1320 V

   ☐ b. 762 V

   ☐ c. 254 V

   ☐ d. 147 V

9. **The primary purpose of a firefighter's switch is to provide:**

☐ a. Isolation

☐ b. Remote control

☐ c. Functional switching

☐ d. Emergency switching

10. **In the formula $Z_s = Z_e + (R_1 + R_2)$ the term Z_s represents the earth fault loop impedance that is:**

☐ a. External to the installation

☐ b. For the combined circuits

☐ c. For the complete system

☐ d. Within the installation

11. **The circuit protective conductors in an electrical installation connect the main earthing terminal to the:**

☐ a. Extraneous conductive parts

☐ b. Exposed conductive parts

☐ c. Metalwork in the building

☐ d. Structural metalwork

12. **The first task when carrying out a risk assessment is to identify:**

☐ a. The risks present

☐ b. The hazards present

☐ c. Who is carrying out the work

☐ d. What PPE equipment is required

13. **A risk assessment should identify if a particular hazard is likely to cause:**

☐ a. Disruption

☐ b. Delays

☐ c. Harm

☐ d. Danger

14. **An employer is required to report an accident to the Health and Safety Executive if an employee is:**

☐ a. Treated at work for a cut hand

☐ b. Sent home for 24 hours due to the accident

☐ c. Unable to work for more than 3 days due to the accident

☐ d. At a hospital casualty department for a morning before returning to work

15. **An Improvement Notice is issued by:**

☐ a. An employee to an employer

☐ b. An HSE inspector

☐ c. A works safety officer

☐ d. An employer to an employee

16. **The term COSHH in the COSHH regulations refers to:**

☐ a. Control of substances hazardous to health

☐ b. Centralizing of substances with health hazards

☐ c. Control of services, health and hazards

☐ d. Control of sustainable health and hygiene

17. **A white cross on a green background identifies the location of the:**

☐ a. Toilet facilities

☐ b. Drying room

☐ c. First aid box

☐ d. Site office

18. **A BS EN 60309 plug for a voltage of 220-240 V is coloured:**

☐ a. Blue

☐ b. Red

☐ c. Green

☐ d. Yellow

19. Wearing ear defenders may be necessary when carrying out maintenance in high ambient noise locations. One problem which may then be encountered is not being able to hear:

☐ a. The radio

☐ b. The mobile phone

☐ c. Any danger warnings

☐ d. Any background noise

20. One relevant statutory document appropriate for electrical maintenance is:

☐ a. The Electricity at Work Regulations

☐ b. The IET Wiring Regulations

☐ c. HSE Guidance Note GS 38

☐ d. The IET Guidance Note 3

21. Which of the following types of trunking would be used to supply sockets mounted at low level in an office building:

☐ a. Skirting

☐ b. Architrave

☐ c. Dado

☐ d. Bench

22. A non-fused spur may be used to connect from a:

☐ a. A socket outlet on a ring to one single-socket outlet

☐ b. Distribution board to two single-socket outlets

☐ c. Junction box connected onto the ring to two twin-socket outlets

☐ d. Consumer unit to any number of single-socket outlets incorporating switches

23. Which of the following cables has both cable screening and pair shielding:

☐ a. U/UTP

☐ b. F/UTP

☐ c. S/FTP

☐ d. SF/UTP

24. The circuit diagram in Figure 1 shows what type of system?

☐ a. Failsafe call system

☐ b. Closed circuit system

☐ c. Push to make system

☐ d. Two detector points in series

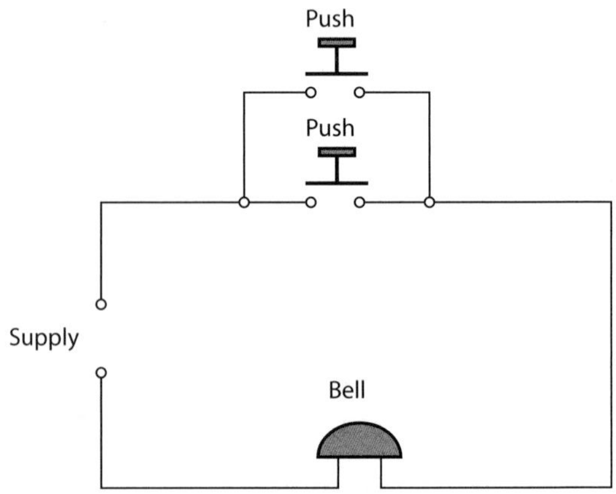

Figure 1

25. The common colour coding method for a typical UTP cable used for a telephone network is:

☐ a. Twisted pair with white foil braiding and solid core colours

☐ b. Four shielded pairs of braid screened cable striped core colours

☐ c. Solid colour twisted pair with foil shielding and braided shielding to the cable

☐ d. Twisted pair with one core solid colour the other core with same colour and white

26. Which of the following is NOT an advantage of planned regular maintenance?

☐ a. Reducing possibility of danger

☐ b. Improving reliability

☐ c. Improving output

☐ d. Reducing costs

27. Records of maintenance should be maintained in order to:

☐ a. Demonstrate compliance with legislation

☐ b. Justify maintenance expenditure

☐ c. Ensure the minimum work is carried out

☐ d. Reduce the number of repairs required

28. The electrical installation in a small industrial unit is subject to regular maintenance and records of these activities are kept. Which of the following statements is correct in relation to this installation?

☐ a. A periodic inspection and test must be carried out every 5 years

☐ b. A periodic inspection and test is required in the event of a breakdown

☐ c. A periodic inspection and test is required annually

☐ d. A periodic inspection and test is not necessary

29. A quality assurance system is employed to ensure that:

☐ a. The same procedure is used for each activity

☐ b. All products are of the very best quality

☐ c. Only certain actions need to be checked

☐ d. Every operative is monitored closely

30. Maintenance records should be made available to the person carrying out the maintenance. Which of the following is NOT one of the items of information that the person performing the maintenance will need to refer to?

☐ a. Previous test results

☐ b. Settings and limits

☐ c. Cost of spare parts

☐ d. Manufacturer's information

Answer section

Chapter 1

Try this Page 6

1 Employers, self-employed and employees
2 a Electric shock
  b Electrical burns
  c Fires of electrical origin
  d Electric arcing
  e Explosions initiated or caused by electricity
3 i The Duty Holder
  ii Records of maintenance, test results and manufacturer's information

Try this Page 11

1 Answers could include, danger from working machines, risk of electric shock, interference from other workers
2 Answers will depend on the dangers identified in 1) Barrier off work areas, safe isolation, warning notices and screens

Task Page 14

1 *Insulation*
2 *Cutting edge*
3 *Teeth*
4 *Pivot (not worn or loose)*
5 *Operation (free to move)*

Task Page 17

Approved Annual Service Schedule
Air Conditioning
Eberspächer

	TASK	ACTION		
1.	Fit refrigerant gauges.			
2.	Check Belt.	Replace belt	Yes / No	
3.	Check Compressor.	Mount kit	Yes / No	
		Tension device	Yes / No	
		Clutch bearing	Yes / No	
		Bolts	Yes / No	
4.	Check hoses.	Fittings	Yes / No	
		Security	Yes / No	
		Chafing	Yes / No	
5.	Check condenser.	Security	Yes / No	
		Clear of debris	Yes / No	
		Fans operating	Yes / No	
6.	Check evaporator.	Security	Yes / No	
		Clear of debris	Yes / No	
		Blower operating	Yes / No	
7.	Pressure switch.	LP		bar
		MP		bar
		HP		bar
8.	Electrics.	Fuses		
		Connections:		
		battery		
		ignition		
		earth		
9.	Controller settings.	Comes		
		Hi	Yes / No	
		Lo	Yes / No	
		Fan speeds	Yes / No	
		Manual thermostat		
		Correct operation	Yes / No	
		Re-circulation controls		
		Operating	Yes / No	
10.	Check for leaks.	Dye added	Yes / No	
11.	Reclaim refrigerant.		Yes	
12.	Replace filter drier.		Yes	
13.	Leak/pressure test system.	20 minutes		bar
14.	Evacuate system.			
15.	Recharge system.	Refrigerant type		
		Refrigerant charge		kg
		Oil type		g
		Oil added		g
16.	Run and test unit.	Pull down to set-point?		
17.	Any additional work.			

Task Page 27

1 *Eye protection*
2 *Work gloves (may include safety shoes and eye protection)*
3 *Ear defenders (may include eye protection)*

SELF ASSESSMENT Page 28

1 a) Any equipment provided by an employer
2 d) Operative's hand tools
3 d) Check for potential hazards in the work area
4 b) Controlled discharge of pressure
5 c) Layout drawing

Chapter 2

Recap Page 29

● statutory persons work
● installation use appropriate working environmental
● inspected dangers risk consideration eliminated
● control
● identify will
● block circuit wiring
● schedules manufacturer followed
● records demonstrates schedule measurements
● Horizontal activities progress
● safely essential
● isolated
● isolated exposed tested dead
● isolating locked off key
● safe us other premises
● Barriers warning advise activity keep danger
● materials site safe secure
● assessed remove reduce danger PPE only alternative

Task Page 34

The equipment required will vary depending on the access method chosen.

*Typically access may be by mobile or fixed scaffolding or a mobile elevated work platform (MEWP). A ladder or steps are **not** acceptable for this purpose.*

The list should also include safety harnesses and anchors together with barriers and warning notices as the work area is used as access by the other trades.

Try this Page 37

Activity 1, between Activities 3 and 4, between Activities 4 and 5, between Activities 5 and 6, Activity 6 and Activity 7

Try this: Crossword Page 42

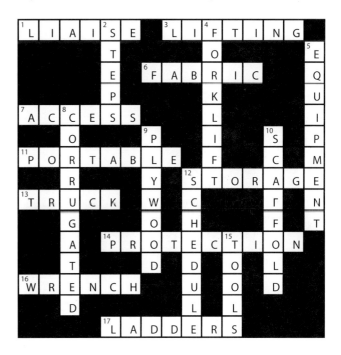

SELF ASSESSMENT Page 43

1 a. Manufacturer's instructions
2 b. Coordinated with other trades
3 d. Sufficient clearance from the structure
4 b. Documented and photographed
5 a. Dust sheets

Chapter 3

Recap Page 44

- confirming   delays   costs
- begins   coordinate   trades   delay
- schedule   information   records   programme
- isolated   battery   task
- centre   reduced   preferred
- only   short
- instruments   calibrated   damage   function
- suitable   safely
- routes   barriers   notices   others
- exit   immediate   damage
- damage   client's   recorded
- protection   structure   fittings
- sheets   plastic   furniture

Try this Page 51

1   a

   b

   c

   d

   e

2   a   Block diagram
   b   Circuit diagram
   c   Wiring diagram
   d   Layout drawing

Try this Page 61

1   Any suitable method of providing the information such as in written format, IT data medium
2   Referring to the appropriate industry and statutory requirements for the fixed installa-tion, the fire alarm and the emergency lighting
3   BS 7671, BS 5839, BS 5266 and IET Electrical Maintenance

Task Page 63

Answers may include

1   The use of pattern parts or reconditioned components
2   The use of LED or low energy lamps
3   Best alternative option is solar-thermal, but biomass or air-source heat pump are other alternatives

SELF ASSESSMENT Page 64

1   a. Wiring diagram
2   b. Reverse conducting diode
3   c. Code of Practice
4   a. NM/1
5   a. Time to complete the work

Chapter 4

Recap Page 65

- client   requirements   statutory   equipment   maintained
- up to date   maintenance
- isolated   operational   each
- advised   cost   adjust   costs   schedule
- components   connected   locations   relation   connections
- proposal   requirements,   particular   industry
- fixed   lighting   equipment   emergency   switchgear
- suitable   friendly   accurate   concise
- work   extent   work   spares

- requirements works environmental personnel health progress

Task Page 72

The answer should include at least:

- *Notification of the delay period*
- *Cause of the delay (manufacturer's supply delayed)*
- *Option to carry out the maintenance to this one machine during the work for days 6 to 8*

Task Page 78

1 BS EN 61558-2-5
2 BS 1363
3 BS 67
4 BS 1362
5 BS EN 60238

Task Page 81

Answer should include:

- Health and Safety at Work (etc.) Act
- Management of Health and Safety at Work Regulations
- Provision and Use of Work Equipment Regulations
- Electricity at Work Regulations
- Workplace (Health Safety and Welfare) Regulations

Task Page 87

1 *Voltage, current, power, physical size and duty*
2 *Answer should include: length, wattage, frequency and colour rendition*

Task Page 91

1 Fibreglass sleeving
2 Cleaning solvent
3 Shellac or varnish

SELF ASSESSMENT Page 92

1 b. The supplier of the work equipment and materials
2 c. Using a variation order
3 c. Non-statutory
4 b. Client's specification
5 a. Type of insulation

End test

1 a. Prior to use
2 a. The date on which work is to be started
3 b. Statement 1 is correct and statement 2 is incorrect
4 b. A permit to work
5 a. Wiring diagram
6 d. Layout diagram
7 b. Bar chart
8 d. The manufacturer's manual
9 a. Gate valve
10 a. 55 V ac
11 d. Laid horizontally on a shelf
12 d. Torque wrench
13 b. Extension ladders
14 d. Calibrated
15 d. Coordinated
16 b. In writing
17 a. Cover them with corrugated plastic
18 a. Variable resistor
19 d. BS 7671 Requirements for Electrical Installations
20 c. Evacuation of the premises
21 b. Annually
22 c. P1
23 d. 15 yearly
24 b) Action ii) only will incur additional cost
25 d. Planned preventative

26 a. Variation order
27 d. Non-routine maintenance
28 b. A current-using device
29 a. A 13 A socket outlet
30 b. Power rating

Chapter 5

Try this Page 102

a Electric shock
b Arcing
c Burns

SELF ASSESSMENT Page 107

1 d) The use of the installation will be restricted
2 a) Increased shock risk
3 a) An approved voltage indicator
4 c) All live conductors and all live conductors and earth
5 b) Working, using a proving unit

Chapter 6

Recap Page 109

- supply off secured re-energized
- isolate equipment risk ourselves
- burns entry exit within path
- persons services
- obtained responsible
  - line neutral
  - line earth
  - Neutral earth
- no present

Try this Page 117

693 V

381 V

208 V

Try this Page 124

1 **a** 400 V ac for heavy plant and equipment, large motors and heating loads, etc.
   b 230 V ac for single phase loads such as socket outlets and lighting
   c 50 V ac ELV for control equipment
   d 12 V ac SELV for controls and equipment in special locations.
2 **a** 230 V
   b 400/230 V
   c 11 kV

Task Page 130

95 A

580 A

520 A

200 A

Try this Page 131

2 The circuit protective conductor; the main earthing terminal; the earthing conductor; the separate metallic conductor in the supply cable; the star point of the supply transformer; the transformer winding; the suppliers line conductor; the circuit line conductor to the point of fault

Try this Crossword Page 132

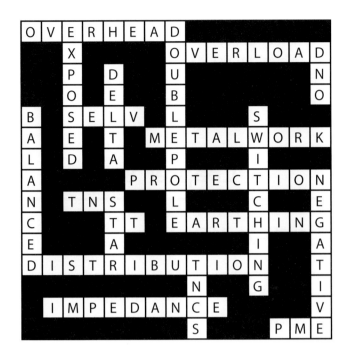

SELF ASSESSMENT Page 133

1 a. TT
2 c. 225 A
3 c. earth leakage currents
4 d. high ac current
5 b. for the complete system

Chapter 7

Recap Page 134

fault   impedance   fault

earthing   PEN   general mass of earth

Secondary   11   Tertiary   400

balanced   neutral   zero

230   single

divided

multiplied

live   short   overload

Try this Page 140

● Working at height: Injury from falls. Potential for material to be dropped from above

● Removal and disposal of fluorescent tubes: Danger from broken lamps due to mercury content

● Working overhead on luminaires above head height: Causing strain on muscles and back

● Lifting materials above head height: Causing strain on muscles and back

Try this Page 144

☑

☑

☒

☒

☑

☒

Try this Page 147

1 a   Control of Substances Hazardous to Health (COSHH) Regulations
   b   Management of Health & Safety at Work Regulations
   c   Personal Protective Equipment at Work Regulations (PPE)
   d   Work at Height Regulations
   e   The Electricity at Work Regulations
   f   Manual Handling Operations Regulations
   g   Workplace (Health and Safety and Welfare) Regulations
   h   Provision and Use of Work Equipment Regulations (PUWER)
2 The employer
3 a   The company's accident book
   b   The Health and Safety Executive (HSE) or the Local Authority Environmental Health Department

Try this Page 151

1 Must include high visibility clothing
2 Should include hard hat, ear defenders, safety shoes possibly safety glasses

Task Page 156

Answer should include:

- Terminal screwdrivers (small and large)
- Large flat blade screwdriver
- Pozidrive screwdrivers (PZ1 and PZ2)
- Electrician's pliers
- Long nose pliers
- Side cutters
- Spanners (various sizes)

Try this Page 159

1 Extension ladder
2 MEWP
3 Step up

Try this Page 159

1 Fibreglass or silicone sleeving
2 Thermoplastic
3 Ceramic
4 Solvent adhesive

Try this Page 163

1 Loss of data
2 Electric shock
3 Falling injury to office staff
4 Damage to the screwdriver handle

SELF ASSESSMENT Page 164

1 b. Hazard
2 c. Safety policy
3 c. When the risk cannot be removed
4 d. 55 V to earth
5 a. using the correct tools

Chapter 8

Recap Page 165

assessment actions reduce carried
potential harm likelihood injury

only last control used

tidy suitable adequate

occurrences deaths enforcing

electric resuscitation sight 24 hours

prohibition stopping immediately appeal

equipment tests compliance

110 V reduced transformer.

lift equipment reform

select equipment duration

unintentional harmful people

functions work damage injury

Task Page 171

Typical Answers:

1 The Health and Safety at Work (etc.) Act and The Electricity at Work Regulations
2 Maintenance schedules and the manufacturer's information

3 Suitable access equipment to enable work on the unit and working overhead

4 Barriers and warning notices to advise and prevent access to the work area

Try this Page 176

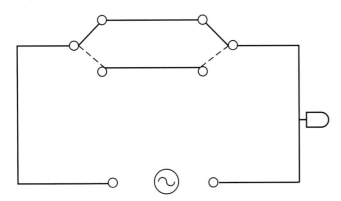

Try this Page 180

Answers could include:

1 **a** Steel wire armoured
   b Skirting or dado trunking
   c Lighting trunking

2 **a** Busbar trunking
   b Steel wire armoured
   c MIMS

Try this Page 183

a

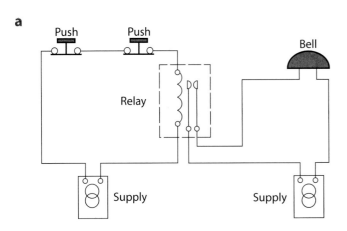

b

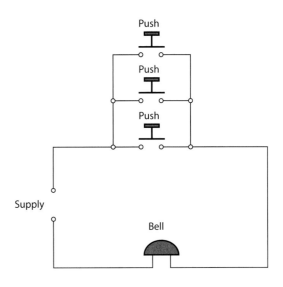

Try this Page 187

a 1) Maintained and 2) non-maintained
b 1) enclosed stairwells with no windows and 2) general fire exits in an office
c FP type cables and MIMS cables

Try this Page 190

1 A UTP cable has no screening/shield whilst a S/STP has screened pairs and an outer cable shield

2 One core solid colour the other core with same colour and white

3 To reflect the light back into the core

Try this Page 195

Answer should include as a minimum:

● *Confirm with client suitable time for the work to be carried out*

● *Consult the manufacturer's information, maintenance schedule, drawings and diagrams*

- *Confirm the necessary spares and equipment are available*
- *Secure the area around the worksite*
- *Position warning notices*
- *Isolate the electrical supply and lock off*

Try this Page 198

Answers could include:

- The cause of the damage (if apparent)
- The requirement for a replacement
- The availablilty of the materials
- The cost and time implications

Task Page 199

Answer should include:

- Isolate and lock off the motor
- Remove the cover from the motor terminal box
- Open the local isolator cover
- Disconnect the conductors at one end and carefully withdraw
- Disconnect and remove the damaged conduit
- Measure, cut and reconnect the new conduit at one end
- Confirm the conductors and insulation are not damaged
- Reinstall the conductors through the conduit
- Terminate the conduit and the conductors
- Confirm correct phase sequence
- Replace covers
- Test motor for correct rotation

Try this Crossword Page 200

SELF ASSESSMENT Page 201

1   c. The manufacturer's instructions
2   d. Two-way and intermediate
3   d. Hold the contacts open
4   c. UTP
5   b. Have a lower power rating

Chapter 9

Recap Page 202

wiring   settings   correct   fault

cpc   subject   adverse   mechanical

trunking   mechanical   single   shared
circuits   rewiring

wiring   flexible   provide   control

busbar   distribution   bar   plug in   power

building   conditioning   fume   ventilation

Closed   primarily

maintained   non-maintained

Area   complex   programmable   building
messages   deal   number   equipment

twisted   telephone   one   solid   colour
white   core   glass   outer   optical   covering

wear   suitable   environmental   operational

procedures   safety   disruption

arrangement   client   shut down

relevant   replacement   decision

accessibility   labour

Try this Page 206

a   Legislative requirements
b   The prevention of danger
c   To reduce costs
d   To maintain reliability
e   Environmental requirements – to prevent
     pollution

Try this Page 213

1   a   Prove that maintenance has taken place
    b   Show the condition of the system and
        equipment
    c   Allow systems and equipment to be
        monitored

2   a   Corrosion of enclosures
    b   Loose electrical connections
    c   Loose fixings
    d   Loose glands or terminations

3   A planned maintenance regime is in opera-
    tion and records of the inspection and test
    results are kept.
4   Any three from the following:
    a   Settings for operational limits
    b   Torque settings for fixings, bolts and
        terminations
    c   Tolerances allowed for bearings, sleeves
        and pistons
    d   Performance characteristics such as fan
        extract rates

SELF ASSESSMENT Page 215

1   b. Increase production
2   b. Statement a) is incorrect and statement b)
    is correct
3   c.   It   demonstrates   compliance   with
    legislation
4   a. Records of test results are kept
5   d.   providing   the   best   possible   service or
    product

End test

1   c) An approved voltage indicator
2   d) 3
3   c) To obtain permission to isolate
4   c) There may be some inconvenience to the
    user
5   d) 400/230 V
6   a) No current at all
7   c) 92 A
8   c) 254 V
9   d) Emergency switching
10 c) For the complete system
11 b) Exposed conductive parts
12 b) The hazards present
13 c) Harm
14 c) Unable to work for more than 3 days due
    to the accident

15 b) An HSE inspector
16 a) Control of substances hazardous to health
17 c) First aid box
18 a) Blue
19 c) Any danger warnings
20 a) The Electricity at Work Regulations
21 a) Skirting
22 a) A socket outlet on a ring to one single-socket outlet
23 c) S/FTP
24 a) Two call points in parallel
25 d) Twisted pair with one core solid colour the other core with same colour and white
26 c) Improving output
27 a) Demonstrate compliance with legislation
28 d) A periodic inspection and test is not necessary
29 a) The same procedure is used for each activity
30 c) Cost of spare parts

Glossary

AVI approved voltage indicator

BS EN British Standard Euro Norm

BS7671 the Requirements for Electrical Installations, is published by the Institution of Engineering and Technology (IET) and is commonly referred to as the Wiring Regulations

BSI The British Standards Institution

Cable type F foil shielding

Cable type S braided shielding

Cable type TP twisted pair

Cable type U unshielded

CAN Controlled Area Network

CENELEC The European Electrical Standards Body

CoP Code of Practice

COSHH Control of Substances Hazardous to Health

cpc circuit protective conductor

csa cross-sectional area

DB Distribution board

DNO Distribution Network Operator

EICR Electrical Installation Condition Reports

ELV extra low voltage

EWR The Electricity at Work Regulations

FP cables which have fire performance rated insulation

Hazard a hazard is anything with the potential to cause harm, so electricity, noise, dust and so on are typical examples of hazards

HSE Health and Safety Executive

HSWA The Health and Safety at Work (etc.) Act

HV high voltage

IET Institution of Engineering and Technology

IP ingress protection

IT a system of electrical supply where there is either no connection of the supply to earth or the system has only a high impedance connection and an insulation monitoring device monitors the impedance. Not used in the public supply network

LAN local area network

LED Light emitting diode

LOLER The Lifting Operations and Lifting Equipment Regulations 1998

LSF low-smoke and fume

LV low voltage

MEWP mobile elevated work platform

MF maintenance free

MICC mineral-insulated copper-clad

MIMS mineral insulated metal sheath cables

OHLS halogen free low smoke

PELV protective extra low voltage

PEN protective earth neutral

PILC paper insulated lead covered

PIR Periodic Inspection Reports

PME protective multiple earthing (see TN-C-S)

PPE personal protective equipment

PUWER Provision and Use of Work Equipment Regulations

PV photovoltaic

QA quality assurance

RCD residual current device

Risk a risk is the likelihood of a hazard causing injury, damage or loss and how severe the outcome may be

SELV separated extra low voltage

SWA steel wire armoured

TN-C the TN-C earthing arrangement is rarely used and it is one where a combined PEN conductor fulfils both the earthing and neutral functions in both the supply and the installation

TN-C-S a supply system in which the earth provision is provided by the DNO using a combined neutral and earth conductor within the supplier's network cables. The earth and neutral are then separated throughout the installation.

These systems are referred to as TN-C-S or PME systems

TN-S a supply system in which the earth provision is provided by the DNO using a separate metallic conductor provided by the DNO. This provision may be by connection to the metal sheath of the supply cable or a separate conductor within the supply cable

TP & N triple-pole and neutral

TT a supply system in which the DNO does not provide an earth facility. The installation's exposed and extraneous metalwork is connected to earth by a separate installation earth electrode and uses the general mass of earth as the return path

VDU visual display unit

XLPE cross-linked polyethylene

Index: EIS book 9